Just The

facts101

Textbook Key Facts

e-Study Guide

Textbook Outlines, Highlights, and Practice Quizzes

Environmental Science: Foundations and Applications

by Andrew Friedland, 1st Edition

All "Just the Facts101" Material Written or Prepared by Cram101 Publishing

Title Page

"Just the Facts101" is a Cram101 publication and tool designed to give you all the facts from your textbooks. Visit Cram101.com for the full practice test for each of your chapters for virtually any of your textbooks.

Cram101 has built custom study tools specific to your textbook. We provide all of the factual testable information and unlike traditional study guides, we will never send you back to your textbook for more information.

YOU WILL NEVER HAVE TO HIGHLIGHT A BOOK AGAIN!

Cram101 StudyGuides

All of the information in this StudyGuide is written specifically for your textbook. We include the key terms, places, people, and concepts... the information you can expect on your next exam!

Want to take a practice test?

Throughout each chapter of this StudyGuide you will find links to cram101.com where you can select specific chapters to take a complete test on, or you can subscribe and get practice tests for up to 12 of your textbooks, along with other exclusive cram101.com tools like problem solving labs and reference libraries.

Cram101.com

Only cram101.com gives you the outlines, highlights, and PRACTICE TESTS specific to your textbook. Cram101.com is an online application where you'll discover study tools designed to make the most of your limited study time.

By purchasing this book, you get 50% off the normal subscription free!. Just enter the promotional code **'DK73DW20190'** on the Cram101.com registration screen.

www.Cram101.com

ISBN(s): 9781478405337. PUBI-2.201279

Learning System

Environmental Science: Foundations and Applications
Andrew Friedland, 1st

CONTENTS

Environmental Science: Studying the State of Our Earth

_____ | Speciation _____

_____ | Air pollution _____

_____ | Pollution _____

_____ | Abiotic component _____

_____ | Biotic component _____

_____ | Environmental science _____

_____ | Atmosphere _____

_____ | CITES _____

_____ | Extinction _____

_____ | Biodiversity _____

_____ | Ecosystem services _____

_____ | Environmental indicator _____

_____ | Sustainability _____

_____ | Species diversity _____

_____ | Tiger _____

_____ | Biodiversity hotspot _____

_____ | Hotspot _____

_____ | Water table _____

_____ | Carbon dioxide _____

Global change

Greenhouse effect

Biome

Natural resource

Resource depletion

Deforestation

Environmental justice

Sustainable development

Ecological footprint

Scientific method

Deductive reasoning

Fly ash

Inductive reasoning

Critical thinking

Environmental mitigation

Chlorpyrifos

Interaction

Subjectivity

Speciation	Speciation is the evolutionary process by which new biological species arise. The biologist Orator F. Cook seems to have been the first to coin the term 'speciation' for the splitting of lineages or 'cladogenesis,' as opposed to 'anagenesis' or 'phyletic evolution' occurring within lineages. Whether genetic drift is a minor or major contributor to speciation is the subject matter of much ongoing discussion.
Air pollution	Air pollution is the introduction of chemicals, particulate matter, or biological materials that cause harm or discomfort to humans or other living organisms, or cause damage to the natural environment or built environment, into the atmosphere.
	The atmosphere is a complex dynamic natural gaseous system that is essential to support life on planet Earth. Stratospheric ozone depletion due to air pollution has long been recognized as a threat to human health as well as to the Earth's ecosystems.
Pollution	Pollution is the introduction of contaminants into a natural environment that causes instability, disorder, harm or discomfort to the ecosystem i.e. physical systems or living organisms. Pollution can take the form of chemical substances or energy, such as noise, heat or light. Pollutants, the components of pollution, can be either foreign substances/energies or naturally occurring contaminants.
Abiotic component	In ecology and biology, abiotic components(also known as abiotic ftors) are non-living chemical and physical ftors in the environment which affect ecosystems. Abiotic phenomena underlie all of biology. Abiotic ftors, while generally downplayed, can have enormous impt on evolution.
Biotic component	Biotic components are the living things that shape an ecosystem. A biotic factor is any living component that affects another organism, including animals that consume the organism in question, and the living food that the organism consumes. Each biotic factor needs energy to do work and food for proper growth.
Environmental science	Environmental science is an interdisciplinary academic field that integrat physical and biological scienc, (including but not limited to Ecology, Physics, Chemistry, Biology, Soil Science, Geology, Atmospheric Science and Geography) to the study of the environment, and the solution of environmental problems. Environmental science provid an integrated, quantitative, and interdisciplinary approach to the study of environmental systems.

	Related areas of study include environmental studi and environmental engineering.
Atmosphere	The standard atmosphere (symbol: atm) is an international reference pressure defined as 101325 Pa and formerly used as unit of pressure. For practical purposes it has been replaced by the bar which is 10^5 Pa. The difference of about 1% is not significant for many applications, and is within the error range of common pressure gges.
CITES	CITES is a multilateral treaty, drafted as a result of a resolution adopted in 1963 at a meeting of members of the International Union for Conservation of Nature (IUCN). The convention was opened for signature in 1973, and CITES entered into force on July 1, 1975. Its aim is to ensure that international trade in specimens of wild animals and plants does not threaten the survival of the species in the wild, and it accords varying degrees of protection to more than 33,000 species of animals and plants. In order to ensure that the General Agreement on Tariffs and Trade (GATT) was not violated, the Secretariat of GATT was consulted during the drafting process.
Extinction	In biology and ecology, extinction is the end of an organism or of a group of organisms (taxon), normally a species. The moment of extinction is generally considered to be the death of the last individual of the species, although the capacity to breed and recover may have been lost before this point. Because a species' potential range may be very large, determining this moment is difficult, and is usually done retrospectively.
Biodiversity	Biodiversity is the degree of variation of life forms within a given species, ecosystem, biome, or an entire planet. Biodiversity is a measure of the health of ecosystems. Biodiversity is in part a function of climate.
Ecosystem services	Humankind benefits from a multitude of resources and processes that are supplied by natural ecosystems. Collectively, these benefits are known as ecosystem services and include products like clean drinking water and processes such as the decomposition of wastes. While scientists and environmentalists have discussed ecosystem services for decades, these services were popularized and their definitions formalized by the United Nations 2004 Millennium Ecosystem Assessment (MA), a four-year study involving more than 1,300 scientists worldwide.

Chapter 1. Environmental Science: Studying the State of Our Earth

Environmental indicator	Environmental indicators are simple measures that tell us what is happening in the environment. Since the environment is very complex, indicators provide a more practical and economical way to track the state of the environment than if we attempted to record every possible variable in the environment. For example, concentrations of ozone depleting substances (ODS) in the atmosphere, tracked over time, is a good indicator with respect to the environmental issue of stratospheric ozone depletion..
Sustainability	Sustainability is the capacity to endure. For humans, sustainability is the long-term maintenance of responsibility, which has environmental, economic, and social dimensions, and encompasses the concept of stewardship, the responsible management of resource use. In ecology, sustainability describes how biological systems remain diverse and productive over time, a necessary precondition for human well-being.
Species diversity	Species diversity is the effective number of different species that are represented in a collection of individuals (a dataset). The effective number of species refers to the number of equally-abundant species needed to obtain the same mean proportional species abundance as that observed in the dataset of interest (where all species may not be equally abundant). Species diversity consists of two components, species richness and species evenness.
Tiger	The tiger is the largest cat species, reaching a total body length of up to 3.3 metres (11 ft) and weighing up to 306 kg (670 lb). Their most recognizable feature is a pattern of dark vertical stripes on reddish-orange fur with lighter underparts. They have exceptionally stout teeth, and their canines are the longest among living felids with a crown height of as much as 74.5 mm (2.93 in) or even 90 mm (3.5 in).
Biodiversity hotspot	A biodiversity hotspot is a biogeographic region with a significant reservoir of biodiversity that is under threat from humans.

The concept of biodiversity hotspots was originated by Norman Myers in two articles in 'The Environmentalist' (1988 ' 1990), revised after thorough analysis by Myers and others in 'Hotspots: Earth's Biologically Richest and Most Endangered Terrestrial Ecoregions'.

To qualify as a biodiversity hotspot on Myers 2000 edition of the hotspot-map, a region must meet two strict criteria: it must contain at least 0.5% or 1,500 species of vascular plants as endemics, and it has to have lost at least 70% of its primary vegetation. |

Hotspot	The places known as hotspots or hot spots in geology are volcanic regions thought to be fed by underlying mantle that is anomalously hot compared with the mantle elsewhere. They may be on, near to, or far from tectonic plate boundaries. There are two hypotheses to explain them.
Water table	The water table is the surface where the water pressure head is equal to the atmospheric pressure (where gauge pressure = 0). It may be conveniently visualized as the 'surface' of the subsurface materials that are saturated with groundwater in a given vicinity. However, saturated conditions may extend above the water table as surface tension holds water in some pores below atmospheric pressure.
Carbon dioxide	Carbon dioxide is a naturally occurring chemical compound composed of two oxygen atoms covalently bonded to a single carbon atom. It is a gas at standard temperature and pressure and exists in Earth's atmosphere in this state, as a trace gas at a concentration of 0.039% by volume. As part of the carbon cycle known as photosynthesis, plants, algae, and cyanobacteria absorb carbon dioxide, light, and water to produce carbohydrate energy for themselves and oxygen as a waste product.
Global change	Global change refers to planetary-scale changes in the Earth system. The system consists of the land, oceans, atmosphere, poles, life, the planet's natural cycles and deep Earth processes. These constituent parts influence one another.
Greenhouse effect	The greenhouse effect is a process by which thermal radiation from a planetary surface is absorbed by atmospheric greenhouse gases, and is re-radiated in all directions. Since part of this re-radiation is back towards the surface, energy is transferred to the surface and the lower atmosphere. As a result, the avera surface temperature is higher than it would be if direct heating by solar radiation were the only warming mechanism.
Biome	Biomes are climatically and geographically defined as similar climatic conditions on the Earth, such as communities of plants, animals, and soil organisms, and are often referred to as ecosystems. Some parts of the earth have more or less the same kind of abiotic and biotic factors spread over a large area, creating a typical ecosystem over that area. Such major ecosystems are termed as biomes.

Chapter 1. Environmental Science: Studying the State of Our Earth

Natural resource	Natural resources occur naturally within environments that exist relatively undisturbed by mankind, in a natural form. A natural resource is often characterized by amounts of biodiversity and geodiversity existent in various ecosystems.
	Natural resources are derived from the environment.
Resource depletion	Resource depletion is an economic term referring to the exhaustion of raw materials within a region. Resources are commonly divided between renewable resources and non-renewable resources. Use of either of these forms of resources beyond their rate of replacement is considered to be resource depletion.
Deforestation	Deforestation is the removal of a forest or stand of trees where the land is thereafter converted to a nonforest use. Examples of deforestation include conversion of forestland to farms, ranches, or urban use.
	The term deforestation is often misused to describe any activity where all trees in an area are removed.
Environmental justice	Environmental justice is 'the fair treatment and meaningful involvement of all people regardless of race, color, sex, national origin, or income with respect to the development, implementation and enforcement of environmental laws, regulations, and policies.' In the words of Bunyan Bryant, 'Environmental justice is served when people can realize their highest potential.'
	Environmental justice emerged as a concept in the United States in the early 1980s; its proponents generally view the environment as encompassing 'where we live, work, and play' (sometimes 'pray' and 'learn' are also included) and seek to redress inequitable distributions of environmental burdens (pollution, industrial facilities, crime, etc).. Root causes of environmental injustices include 'institutionalized racism; the co-modification of land, water, energy and air; unresponsive, unaccountable government policies and regulation; and lack of resources and power in affected communities.'
	Definition

	The United States Environmental Protection Agency defines as follows:
	'Environmental Justice is the fair treatment and meaningful involvement of all people regardless of race, color, national origin, or income with respect to the development, implementation, and enforcement of environmental laws, regulations, and policies. EPA has this goal for all communities and persons across this Nation.
Sustainable development	Sustainable development is a pattern of growth in which resource use aims to meet human needs while preserving the environment so that these needs can be met not only in the present, but also for generations to come (sometimes taught as ELF-Environment, Local people, Future). The term sustainable development was used by the Brundtland Commission which coined what has become the most often-quoted definition of sustainable development as development that 'meets the needs of the present without compromising the ability of future generations to meet their own needs.'
	Sustainable development ties together concern for the carrying capacity of natural systems with the social challenges facing humanity. As early as the 1970s 'sustainability' was employed to describe an economy 'in equilibrium with basic ecological support systems.' Ecologists have pointed to The Limits to Growth, and presented the alternative of a 'steady state economy' in order to address environmental concerns.
Ecological footprint	The ecological footprint is a measure of human demand on the Earth's ecosystems. It is a standardized measure of demand for natural capital that may be contrasted with the planet's ecological capacity to regenerate. It represents the amount of biologically productive land and sea area necessary to supply the resources a human population consumes, and to assimilate associated waste.

Chapter 1. Environmental Science: Studying the State of Our Earth

Scientific method	Scientific method refers to a body of techniques for investigating phenomena, acquiring new knowledge, or correcting and integrating previous knowledge. To be termed scientific, a method of inquiry must be based on gathering empirical and measurable evidence subject to specific principles of reasoning. The Oxford English Dictionary says that scientific method is: 'a method or procedure that has characterized natural science since the 17th century, consisting in systematic observation, measurement, and experiment, and the formulation, testing, and modification of hypotheses.'
	The chief characteristic which distinguishes a scientific method of inquiry from other methods of acquiring knowledge is that scientists seek to let reality speak for itself, and contradict their theories about it when those theories are incorrect, i. e., falsifiability.
Deductive reasoning	Deductive reasoning, is reasoning which constructs or evaluates deductive arguments. Deductive reasoning contrasts with inductive reasoning in that a specific conclusion is arrived at from a general principle. Deductive arguments are attempts to show that a conclusion necessarily follows from a set of premises or hypotheses.
Fly ash	Fly ash is one of the residues generated in combustion, and comprises the fine particles that rise with the flue gases. Ash which does not rise is termed bottom ash. In an industrial context, fly ash usually refers to ash produced during combustion of coal.
Inductive reasoning	Inductive reasoning, is a kind of reasoning that constructs or evaluates propositions that are abstractions of observations of individual instances of members of the same class. Inductive reasoning contrasts with deductive reasoning in that a general conclusion is arrived at by specific examples.
	Definition of inductive reasoning However, philosophically the definition is much more nuanced than simple progression from particular / individual instances to wider generalizations.
Critical thinking	Critical thinking is the process of thinking that questions assumptions. It is a way of deciding whether a claim is true, false; sometimes true, or partly true. The origins of critical thinking can be traced in Western thought to the Socratic method of Ancient Greece and in the East, to the Buddhist kalama sutta and Abhidharma.

Environmental mitigation	Environmental mitigation, compensatory mitigation, or mitigation banking, are terms used primarily by the United States government and the related environmental industry to describe projects or programs intended to offset known impacts to an existing historic or natural resource such as a stream, wetland, endangered species, archeological site or historic structure. To 'mitigate' means to make less harsh or hostile. Environmental mitigation is typically a part of an environmental crediting syst established by governing bodies which involves allocating debits and credits.
Chlorpyrifos	Chlorpyrifos is a crystalline organophosphate insecticide that inhibits acetylcholinesterase and is used to control insect pests. Trade names include Brodan, Detmol UA, Dowco 179, Dursban, Empire, Eradex, Lorsban, Paqeant, Piridane, Scout, and Stipend. Chlorpyrifos is moderately toxic and chronic exposure has been linked to neurological effects, developmental disorders, and autoimmune disorders.
Interaction	Interaction is a kind of action that occurs as two or more objects have an effect upon one another. The idea of a two-way effect is essential in the concept of interaction, as opposed to a one-way causal effect. A closely related term is interconnectivity, which deals with the interactions of interactions within systems: combinations of many simple interactions can lead to surprising emergent phenomena.
Subjectivity	Subjectivity refers to the subject and his or her perspective, feelings, beliefs, and desires. In philosophy, the term is usually contrasted with objectivity. Qualia Subjectivity may refer to the specific discerning interpretations of any aspect of experiences.

---

1. _____ is the introduction of chemicals, particulate matter, or biological materials that cause harm or discomfort to humans or other living organisms, or cause damage to the natural environment or built environment, into the atmosphere.

   The atmosphere is a complex dynamic natural gaseous system that is essential to support life on planet Earth. Stratospheric ozone depletion due to _____ has long been recognized as a threat to human health as well as to the Earth's ecosystems.

   a. Aerotoxic Association
   b. Aerotoxic syndrome
   c. Air pollution
   d. Air quality

2. _____ is the evolutionary process by which new biological species arise. The biologist Orator F. Cook seems to have been the first to coin the term '_____' for the splitting of lineages or 'cladogenesis,' as opposed to 'anagenesis' or 'phyletic evolution' occurring within lineages. Whether genetic drift is a minor or major contributor to _____ is the subject matter of much ongoing discussion.

   a. Polyploid
   b. Juglone
   c. Gibbons v. Ogden
   d. Speciation

3. _____ is the introduction of contaminants into a natural environment that causes instability, disorder, harm or discomfort to the ecosystem i.e. physical systems or living organisms. _____ can take the form of chemical substances or energy, such as noise, heat or light. Pollutants, the components of _____, can be either foreign substances/energies or naturally occurring contaminants.

   a. Basic precipitation
   b. Bioaccumulation
   c. Pollution
   d. Biofilter

4. _____s are the living things that shape an ecosystem. A biotic factor is any living component that affects another organism, including animals that consume the organism in question, and the living food that the organism consumes. Each biotic factor needs energy to do work and food for proper growth.

---

a. Carolina Bay
b. Carrion
c. Biotic component
d. Chreod

5. The standard _____(symbol: atm) is an international reference pressure defined as 101325 Pa and formerly used as unit of pressure. For practical purposes it has been replaced by the bar which is 10^5 Pa. The difference of about 1% is not significant for many applications, and is within the error range of common pressure gges.

a. Atmospheric dispersion modeling
b. Earth systems engineering and management
c. Ecological sanitation
d. Atmosphere

---

1. c

2. d

3. c

4. c

5. d

You can take the complete Chapter Practice Test

for Chapter 1. Environmental Science: Studying the State of Our Earth
on all key terms, persons, places, and concepts.

Online 99 Cents

http://www.epub89.16.20190.1.cram101.com/

Use www.Cram101.com for all your study needs

including Cram101's online interactive problem solving labs in chemistry, statistics, mathematics, and more.

Environmental Systems: Matter, Energy, and Change

	Invertebrate
	Alkali
	Carrying capacity
	Atomic number
	Electron
	Periodic table
	Proton
	Isotope
	Mass number
	Radioactive decay
	Chemical bond
	Hydrogen bond
	Ionic bond
	Methane
	Sodium chloride
	Covalent bond
	Greenhouse gas
	Capillary action
	Surface tension

_____ | Molecule ____

_____ | Nitric acid ____

_____ | Solvent ____

_____ | Sulfuric acid ____

_____ | Calcium hydroxide ____

_____ | Chemical reaction ____

_____ | Nuclear reaction ____

_____ | Sodium hydroxide ____

_____ | Air pollution ____

_____ | Hydroxide ____

_____ | Pollution ____

_____ | Ammonia ____

_____ | Carbon dioxide ____

_____ | Enzyme ____

_____ | Global change ____

_____ | Green Revolution ____

_____ | Macromolecule ____

_____ | Monosaccharide ____

_____ | Polysaccharide ____

Chapter 2. Environmental Systems: Matter, Energy, and Change

_____ | Solar energy _____

_____ | Steppe _____

_____ | Photon _____

_____ | Ethanol _____

_____ | Radiation _____

_____ | Kinetic energy _____

_____ | Potential _____

_____ | Potential energy _____

_____ | Chemical energy _____

_____ | Second law of thermodynamics _____

_____ | Thermodynamics _____

_____ | Efficiency _____

_____ | Nuclear power _____

_____ | Energy quality _____

_____ | Fly ash _____

_____ | Tundra _____

_____ | Closed system _____

_____ | Open system _____

_____ | Systems analysis _____

Energy flow

Steady state

Negative feedback

Positive feedback

Global warming

Ice age

Biome

Sustainability

Phosphorus cycle

Sustainable management

Wetland

Adaptive management

Invasive species

Chlorofluorocarbon

Ozone

Ozone layer

Catalyst

Environmental mitigation

Montreal Protocol

Invertebrate	An invertebrate is an animal without a backbone. The group includes 97% of all animal species - all animals except those in the chordate subphylum Vertebrata (fish, amphibians, reptiles, birds, and mammals).
	Invertebrates form a paraphyletic group.
Alkali	In chemistry, an alkali is a basic, ionic salt of an alkali metal or alkaline earth metal element. Some authors also define an alkali as a base that dissolves in water. A solution of a soluble base has a pH greater than 7. The adjective alkaline is commonly used in English as a synonym for base, especially for soluble bases.
Carrying capacity	The carrying capacity of a biological species in an environment is the maximum population size of the species that the environment can sustain indefinitely, given the food, habitat, water and other necessities available in the environment. In population biology, carrying capacity is defined as the environment's maximal load, which is different from the concept of population equilibrium.
	For the human population, more complex variables such as sanitation and medical care are sometimes considered as part of the necessary establishment.
Atomic number	In chemistry d physics, the atomic number is the number of protons found in the nucleus of atom d therefore identical to the charge number of the nucleus. It is conventionally represented by the symbol Z. The atomic number uniquely identifies a chemical element. In atom of neutral charge, the atomic number is also equal to the number of electrons.
Electron	The electron is a subatomic particle with a negative elementary electric charge. It has no known components or substructure; in other words, it is generally thought to be an elementary particle. An electron has a mass that is approximately 1/1836 that of the proton.

Periodic table	The periodic table is a tabular display organizing the 118 known chemical elements by selected properties of their atomic structure. Elements are presented in the periodic table by increasing values of their atomic numbers, the number of protons in their atomic nuclei. While rectangular in general outline, counter-intuitive gaps are included in the horizontal rows ('periods') as needed to keep elements with similar properties together in each vertical column ('group'), such as the alkali metals, the alkali earths, the halogens, and the noble gases.
Proton	The proton is a subatomic particle with the symbol p or p^+ and a positive electric charge of 1 elementary charge. One or more protons are present in the nucleus of each atom, along with neutrons. The number of protons in each atom is its atomic number.
Isotope	Isotopes are variants of a particular chemical element. While all isotopes of a given element share the same number of protons, each isotope differs from the others in its number of neutrons. The term isotope is formed from the Greek roots isos (?σος 'equal') and topos (τ?πος 'place').
Mass number	The mass number also called atomic mass number, is the total number of protons and neutrons (together known as nucleons) in an atomic nucleus. Because protons and neutrons both are baryons, the mass number A is identical with the baryon number B as of the nucleus as of the whole atom or ion. The mass number is different for each different isotope of a chemical element.
Radioactive decay	Radioactive decay is the process by which an atomic nucleus of an unstable atom loses energy by emitting ionizing particles (ionizing radiation) A decay, or loss of energy, results when an atom with one type of nucleus, called the parent radionuclide, transforms to an atom with a nucleus in a different state, or to a different nucleus containing different numbers of nucleons.
Chemical bond	A chemical bond is an attraction between atoms that allows the formation of chemical substances that contain two or more atoms. The bond is caused by the electromagnetic force attraction between opposite charges, either between electrons and nuclei, or as the result of a dipole attraction. The strength of chemical bonds varies considerably; there are 'strong bonds' such as covalent or ionic bonds and 'weak bonds' such as dipole-dipole interactions, the London dispersion force and hydrogen bonding.
Hydrogen bond	A hydrogen bond is the attractive interaction of a hydrogen atom with an electronegative atom, such as nitrogen, oxygen or fluorine, that comes from another molecule or chemical group. The hydrogen must be covalently bonded to another electronegative atom to create the bond. These bonds can occur between molecules (intermolecularly), or within different parts of a single molecule (intramolecularly).

Chapter 2. Environmental Systems: Matter, Energy, and Change

Ionic bond	An ionic bond is a type of chemical bond formed through an electrostatic attraction between two oppositely charged ions. Ionic bonds are formed between a cation, which is usually a metal, and an anion, which is usually a nonmetal. Pure ionic bonding cannot exist: all ionic compounds have some degree of covalent bonding.
Methane	Appendix: extraterrestrial methane Methane has been detected or is believed to exist in several locations of the solar system. In most cases, it is believed to have been created by abiotic processes. Possible exceptions are Mars and Titan.
Sodium chloride	Sodium chloride, common salt, table salt or halite, is an ionic compound with the formula NaCl. Sodium chloride is the salt most responsible for the salinity of the ocean and of the extracellular fluid of many multicellular organisms. As the major ingredient in edible salt, it is commonly used as a condiment and food preservative.
Covalent bond	A covalent bond is a form of chemical bonding that is characterized by the sharing of pairs of electrons between atoms. The stable balance of attractive and repulsive forces between atoms when they share electrons is known as covalent bonding. Covalent bonding includes many kinds of interaction, including σ-bonding, π-bonding, metal-to-metal bonding, agostic interactions, and three-center two-electron bonds.
Greenhouse gas	A greenhouse gas is a gas in an atmosphere that absorbs and emits radiation within the thermal infrared range. This process is the fundamental cause of the greenhouse effect. The primary greenhouse gases in the Earth's atmosphere are water vapor, carbon dioxide, methane, nitrous oxide, and ozone.
Capillary action	Capillary action, is the ability of a liquid to flow against gravity where liquid spontaneously rises in a narrow space such as a thin tube, or in porous materials such as paper or in some non-porous materials such as liquified carbon fibre. This effect can cause liquids to flow against the force of gravity or the magnetic field induction. It occurs because of inter-molecular attractive forces between the liquid and solid surrounding surfaces; If the diameter of the tube is sufficiently small, then the combination of surface tension (which is caused by cohesion within the liquid) and forces of adhesion between the liquid and container act to lift the liquid.

Surface tension	Surface tension is a property of the surface of a liquid that allows it to resi an external force. It is revealed, for example, in the floating of some objects on the surface of water, even though they are denser than water, and in the ability of some insects (e.g. water riders) to run on the water surface. This property is caused by cohesion of similar molecules, and is responsible for many of the behaviors of liquids.
Molecule	A molecule is an electrically neutral group of two or more atoms held together by covalent chemical bonds. Molecules are distinguished from ions by their electrical charge. However, in quantum physics, organic chemistry, and biochemistry, the term molecule is often used less strictly, also being applied to polyatomic ions.
Nitric acid	Nitric acid also known as aqua fortis and spirit of nitre, is a highly corrosive and toxic strong mineral acid which is normally colorless but tends to acquire a yellow cast due to the accumulation of oxides of nitrogen if long-stored. Ordiry nitric acid has a concentration of 68%. When the solution contains more than 86% HNO_3, it is referred to as fuming nitric acid.
Solvent	A solvent is a liquid, solid, or gas that dissolves another solid, liquid, or gaseous solute, resulting in a solution that is soluble in a certain volume of solvent at a specified temperature. Common uses for organic solvents are in dry cleaning (e.g., tetrachloroethylene), as a paint thinner (e.g., toluene, turpentine), as nail polish removers and glue solvents (acetone, methyl acetate, ethyl acetate), in spot removers (e.g., hexane, petrol ether), in detergents (citrus terpenes), in perfumes (ethanol), nail polish and in chemical synthesis. The use of inorganic solvents (other than water) is typically limited to research chemistry and some technological processes.
Sulfuric acid	Sulfuric acid is a highly corrosive strong mineral acid with the molecular formula H_2SO_4. The historical name of this acid is oil of vitriol. It is a colorless to slightly yellow viscous liquid and is soluble in water at all concentrations.
Calcium hydroxide	Calcium hydroxide, traditionally called slaked lime, is an inorganic compound with the emical formula $Ca(OH)_2$. It is a colorless crystal or white powder and is obtained when calcium oxide (called lime or quicklime) is mixed, or 'slaked' with water. It has many names including hydrated lime, builders lime, slack lime, cal, or pickling lime.

Chapter 2. Environmental Systems: Matter, Energy, and Change

Chemical reaction	A chemical reaction is a process that leads to the transformation of one set of chemical substances to another. Chemical reactions can be either spontaneous, requiring no input of energy, or non-spontaneous, typically following the input of some type of energy, such as heat, light or electricity. Classically, chemical reactions encompass changes that strictly involve the motion of electrons in the forming and breaking of chemical bonds, although the general concept of a chemical reaction, in particular the notion of a chemical equation, is applicable to transformations of elementary particles (such as illustrated by Feynman diagrams), as well as nuclear reactions.
Nuclear reaction	In nuclear physics and nuclear chemistry, a nuclear reaction is semantically considered to be the process in which two nuclei, or else a nucleus of an atom and a subatomic particle (such as a proton, neutron, or high energy electron) from outside the atom, collide to produce products different from the initial particles. In principle, a reaction can involve more than three particles colliding, but because the probability of three or more nuclei to meet at the same time at the same place is much less than for two nuclei, such an event is exceptionally rare. Radioactive decays can be considered to be spontaneous nuclear reactions, in as much as there is such a thing as a spontaneous chemical reaction.
Sodium hydroxide	Sodium hydroxide also known as lye and caustic soda, is a caustic metallic base. It is used in many industries, mostly as a strong chemical base in the manufacture of pulp and paper, textiles, drinking water, soaps and detergents and as a drain cleaner. Worldwide production in 2004 was approximately 60 million tonnes, while demand was 51 million tonnes.
Air pollution	Air pollution is the introduction of chemicals, particulate matter, or biological materials that cause harm or discomfort to humans or other living organisms, or cause damage to the natural environment or built environment, into the atmosphere.
	The atmosphere is a complex dynamic natural gaseous system that is essential to support life on planet Earth. Stratospheric ozone depletion due to air pollution has long been recognized as a threat to human health as well as to the Earth's ecosystems.
Hydroxide	Hydroxide is a diatomic anion with chemical formula OH^-. It consists of an oxygen and a hydrogen atom held together by a covalent bond, and carrying a negative electric charge. It is an important but usually minor constituent of water.

Pollution	Pollution is the introduction of contaminants into a natural environment that causes instability, disorder, harm or discomfort to the ecosystem i.e. physical systems or living organisms. Pollution can take the form of chemical substances or energy, such as noise, heat or light. Pollutants, the components of pollution, can be either foreign substances/energies or naturally occurring contaminants.
Ammonia	Ammonia is a compound of nitrogen and hydrogen with the formula NH_3. It is a colourless gas with a characteristic pungent odor. Ammonia contributes significantly to the nutritional needs of terrestrial organisms by serving as a precursor to food and fertilizers.
Carbon dioxide	Carbon dioxide is a naturally occurring chemical compound composed of two oxygen atoms covalently bonded to a single carbon atom. It is a gas at standard temperature and pressure and exists in Earth's atmosphere in this state, as a trace gas at a concentration of 0.039% by volume. As part of the carbon cycle known as photosynthesis, plants, algae, and cyanobacteria absorb carbon dioxide, light, and water to produce carbohydrate energy for themselves and oxygen as a waste product.
Enzyme	Enzymes ( ) are proteins that catalyze (i.e., increase the rates of) chemical reactions. In enzymatic reactions, the molecules at the beginning of the process, called substrates, are converted into different molecules, called products. Almost all chemical reactions in a biological cell need enzymes in order to occur at rates sufficient for life.
Global change	Global change refers to planetary-scale changes in the Earth system. The system consists of the land, oceans, atmosphere, poles, life, the planet's natural cycles and deep Earth processes. These constituent parts influence one another.
Green Revolution	Green Revolution refers to a series of research, development, and technology transfer initiatives, occurring between the 1940s and the late 1970s, that increased agriculture production around the world, beginning most markedly in the late 1960s.

The initiatives, led by Norman Borlaug, the 'Father of the Green Revolution' credited with saving over a billion people from starvation, involved the development of high-yielding varieties of cereal grains, expansion of irrigation infrastructure, modernization of management techniques, distribution of hybridized seeds, synthetic fertilizers, and pesticides to farmers.

The term 'Green Revolution' was first used in 1968 by former United States Agency for International Development (USAID) director William Gaud, who noted the spread of the new technologies and said,

These and other developments in the field of agriculture contain the makings of a new revolution.

Macromolecule	A macromolecule is a very large molecule commonly created by polymerization of smaller subunits. In biochemistry, the term is applied to the four conventional biopolymers (nucleic acids, proteins, carbohydrates, and lipids), as well as non-polymeric molecules with large molecular mass such as macrocycles. The individual constituent molecules of macromolecules are called monomers (mono=single, meros=part).
Monosaccharide	Monosaccharides are the most basic units of biologically important carbohydrates. They are the simplest form of sugar and are usually colorless, water-soluble, crystalline solids. Some monosaccharides have a sweet taste.
Polysaccharide	Polysaccharides are long carbohydrate molecules of repeated monomer units joined together by glycosidic bonds. They range in structure from linear to highly branched. Polysaccharides are often quite heterogeneous, containing slight modifications of the repeating unit.
Solar energy	Solar energy, radiant light and heat from the sun, has been harnessed by humans since ancient times using a range of ever-evolving technologies. Solar radiation, along with secondary solar-powered resources such as wind and wave power, hydroelectricity and biomass, account for most of the available renewable energy on earth. Only a minuscule fraction of the available solar energy is used.

Steppe	In physical geography, a steppe is an ecoregion, in the montane grasslands and shrublands and temperate grasslands, savannas, and shrublands biomes, characterized by grassland plains without trees apart from those near rivers and lakes. The prairie (especially the shortgrass and mixed prairie) is an example of a steppe, though it is not usually called such. It may be semi-desert, or covered with grass or shrubs or both, depending on the season and latitude.
Photon	In physics, a photon is an elementary particle, the quantum of light and all other forms of electromagnetic radiation, and the force carrier for the electromagnetic force. The effects of this force are easily observable at both the microscopic and macroscopic level, because the photon has no rest mass; this allows for interactions at long distances. Like all elementary particles, photons are currently best explained by quantum mechanics and exhibit wave-particle duality, exhibiting properties of both waves and particles.
Ethanol	Ethanol, pure alcohol, grain alcohol, or drinking alcohol, is a volatile, flammable, colorless liquid. It is a psychoactive drug and one of the oldest recreational drugs. Best known as the type of alcohol found in alcoholic beverages, it is also used in thermometers, as a solvent, and as a fuel.
Radiation	In physics, radiation is a process in which energetic particles or energetic waves travel through a medium or space. Two types of radiation are commonly differentiated in the way they interact with normal chemical matter: ionizing and non-ionizing radiation. The word radiation is often colloquially used in reference to ionizing radiation but the term radiation may correctly also refer to non-ionizing radiation.
Kinetic energy	The kinetic energy of an object is the energy which it possesses due to its motion. It is defined as the work needed to accelerate a body of a given mass from rest to its stated velocity. Having gained this energy during its acceleration, the body maintains this kinetic energy unless its speed changes.
Potential	<ul><li>In linguistics, the potential mood</li><li>The mathematical study of potentials is known as potential theory; it is the study of harmonic functions on manifolds. This mathematical formulation arises from the fact that, in physics, the scalar potential is irrotational, and thus has a vanishing Laplacian -- the very definition of a harmonic function.</li><li>In physics, a potential may refer to the scalar potential or to the vector potential. In either case, it is a field defined in space, from which many important physical properties may be derived.</li></ul>

Chapter 2. Environmental Systems: Matter, Energy, and Change

Potential energy	In physics, potential energy is the energy of a body or a system due to the position of the body or the arrangement of the particles of the system. The SI unit for measuring work and energy is the Joule (symbol J). The term 'potential energy' was coined by the 19th century Scottish engineer and physicist William Rankine.
Chemical energy	In chemistry, Chemical energy is the potential of a chemical substan to undergo a transformation through a chemical reaction or, to transform other chemical substans.Template:Fusion Breaking or making of chemical bonds involves energy, which may be either absorbed or evolved from a chemical system. Energy that can be released (or absorbed) because of a reaction between a set of chemical substans is equal to the differen between the energy content of the products and the reactants. This change in energy is change in internal energy of a chemical reaction.
Second law of thermodynamics	The second law of thermodynamics is an expression of the tendency that over time, differences in temperature, pressure, and chemical potential equilibrate in an isolated physical system. From the state of thermodynamic equilibrium, the law deduced the principle of the increase of entropy and explains the phenomenon of irreversibility in nature. The second law declares the impossibility of machines that generate usable energy from the abundant internal energy of nature by processes called perpetual motion of the second kind.
Thermodynamics	ImgProperty database img Thermodynamics is the branch of physical science concerned with heat and its relation to other forms of energy and work. It defines macroscopic variables (such as temperature, entropy, and pressure) that describe average properties of material bodies and radiation, and explains how they are related and by what laws they change with time. Thermodynamics does not describe the microscopic constituents of matter, and its laws can be derived from statistical mechanics.
Efficiency	Efficiency in general describes the extent to which time or effort is well used for the intended task or purpose. It is often used with the specific purpose of relaying the capability of a specific application of effort to produce a specific outcome effectively with a minimum amount or quantity of waste, expense, or unnecessary effort. 'Efficiency' has widely varying meanings in different disciplines.

Nuclear power	Nuclear power is the use of sustained nuclear fission to generate heat and electricity. Nuclear power plants provide about 6% of the world's energy and 13-14% of the world's electricity, with the U.S., France, and Japan together accounting for about 50% of nuclear generated electricity. In 2007, the IAEA reported there were 439 nuclear power reactors in operation in the world, operating in 31 countries.
Energy quality	Energy quality is the contrast between different forms of energy, the different trophic levels in ecological systems and the propensity of energy to convert from one form to another. The concept refers to the empirical experience of the characteristics, or qualia, of different energy forms as they flow and transform. It appeals to our common perception of the heat value, versatility, and environmental performance of different energy forms and the way a small increment in energy flow can sometimes produce a large transformation effect on both energy physical state and energy.
Fly ash	Fly ash is one of the residues generated in combustion, and comprises the fine particles that rise with the flue gases. Ash which does not rise is termed bottom ash. In an industrial context, fly ash usually refers to ash produced during combustion of coal.
Tundra	In physical geography, tundra is a biome where the tree growth is hindered by low temperatures and short growing seasons. The term tundra comes through Russian тундра from the Kildin Sami word tundâr 'uplands,' 'treeless mountain tract.' There are three types of tundra: Arctic tundra, alpine tundra, and Antarctic tundra. In tundra, the vegetation is composed of dwarf shrubs, sedges and grasses, mosses, and lichens.
Closed system	The term closed system has different meanings in different contexts. In thermodynami In thermodynami, a closed system can exchange energy (as heat or work), but not matter, with its surroundings. In contrast, an isolated system cannot exchange any of heat, work, or matter with the surroundings, while an open system can exchange all of heat, work and matter.
Open system	An open system is a system which continuously interacts with its environment. An open system should be contrasted with the concept of an isolated system which exchanges neither energy, matter,nor information with its environment.

The concept of an 'open system' was formalized within a framework that enabled one to interrelate the theory of the organism, thermodynamics, and evolutionary theory.

Systems analysis	Systems analysis is the study of sets of interacting entities, including computer systems analysis. This field is closely related to requirements analysis or operations research. It is also 'an explicit formal inquiry carried out to help someone (referred to as the decision maker) identify a better course of action and make a better decision than he might otherwise have made.'

Overview
The terms analysis and synthesis come from Greek where they mean respectively 'to take apart' and 'to put together'.

Energy flow	In ecology, energy flow, also called the calorific flow, rrs to the flow of energy through a food chain. In an ecosystem, ecologists seek to quantify the relative importance of different component species and feeding relationships.

A general energy flow scenario follows:

- Solar energy is fixed by the photoautotrophs, called primary producers, like green plants.

Steady state	A system in a steady state has numerous properties that are unchanging in time. This implies that for any property p of the system, the partial derivative with respect to time is zero:

$$\frac{\partial p}{\partial t} = 0$$

The concept of steady state has relevance in many fields, in particular thermodynamics and economics. Steady state is a more general situation than dynamic equilibrium.

Negative feedback	Negative feedback occurs when the output of a system acts to oppose changes to the input of the system, with the result that the changes are attenuated. If the overall feedback of the system is negative, then the system will tend to be stable. Overview In many physical and biological systems, qualitatively different iluences can oppose each other.
Positive feedback	Positive feedback is a process in which the effects of a small disturbance on (a perturbation of) a system include an increase in the magnitude of the perturbation. That is, A produces more of B which in turn produces more of A. In contrast, a system that responds to a perturbation in a way that reduces its effect is said to exhibit negative feedback. These concepts were first recognized as broadly applicable by Norbert Wiener in his 1948 work on cybernetics.
Global warming	Global warming refers to the rising average temperature of Earth's atmosphere and oceans, which began to increase in the late 19th century and is projected to continue rising. Since the early 20th century, Earth's average surface temperature has increased by about 0.8 °C (1.4 °F), with about two thirds of the increase occurring since 1980. Warming of the climate system is unequivocal, and scientists are more than 90% certain that most of it is caused by increasing concentrations of greenhouse gases produced by human activities such as deforestation and the burning of fossil fuels. These findings are recognized by the national science academies of all major industrialized nations.[A] Climate model projections are summarized in the 2007 Fourth Assessment Report (AR4) by the Intergovernmental Panel on Climate Change (IPCC).
Ice age	An ice age, a glacl age, is a period of long-term reduction in the temperature of the Earth's surface and atmosphere, resulting in the presence or expansion of continental ice sheets, polar ice sheets and alpine glaciers. Within a long-term ice age, individual pulses of cold climate are termed 'glacl periods' (or alternatively 'glacls' or 'glactions' or colloqully as 'ice age'), and intermittent warm periods are called 'interglacls'. Glaciologically, ice age implies the presence of extensive ice sheets in the northern and southern hemispheres.

Chapter 2. Environmental Systems: Matter, Energy, and Change

Biome	Biomes are climatically and geographically defined as similar climatic conditions on the Earth, such as communities of plants, animals, and soil organisms, and are often referred to as ecosystems. Some parts of the earth have more or less the same kind of abiotic and biotic factors spread over a large area, creating a typical ecosystem over that area. Such major ecosystems are termed as biomes.
Sustainability	Sustainability is the capacity to endure. For humans, sustainability is the long-term maintenance of responsibility, which has environmental, economic, and social dimensions, and encompasses the concept of stewardship, the responsible management of resource use. In ecology, sustainability describes how biological systems remain diverse and productive over time, a necessary precondition for human well-being.
Phosphorus cycle	The phosphorus cycle is the biogeochemical cycle that describes the movement of phosphorus through the lithosphere, hydrosphere, and biosphere. Unlike many other biogeochemical cycles, the atmosphere does not play a significant role in the movement of phosphorus, because phosphorus and phosphorus-based compounds are usually solids at the typical ranges of temperature and pressure found on Earth. The production of phosphine gas occurs only in specialized, local conditions.
Sustainable management	Sustainable management takes the concepts from sustainability and synthesizes them with the concepts of management. Sustainability has three branches: the environment, the needs of present and future generations, and the economy. Using these branches, it creates the ability to keep a system running indefinitely without depleting resources, maintaining economic viability, and also nourishing the needs of the present and future generations.
Wetland	A wetland is a land area that is saturated with water, either permanently or seasonally, such that it takes on characteristics that distinguish it as a distinct ecosystem. The primary factor that distinguishes wetlands is the characteristic vegetation that is adapted to its unique soil conditions: Wetlands are made up primarily of hydric soil, which supports aquatic plants. The water found in wetlands can be saltwater, freshwater, or brackish.
Adaptive management	What is Adaptive Management ?

Adaptive management also known as adaptive resource management (ARM), is a structured, iterative process of optimal decision making in the face of uncertainty, with an aim to reducing uncertainty over time via system monitoring. In this way, decision making simultaneously maximizes one or more resource objectives and, either passively or actively, accrues information needed to improve future management. Adaptive management is a tool which should be used not only to change a system, but also to learn about the system (Holling 1978).

Invasive species	Invasive species, a nomenclature term and categorization phrase used for flora and fauna, and for specific restoration-preservation processes in native habitats, with several definitions.

- The first definition, the most used, applies to introduced species (also called 'non-indigenous' or 'non-native') that adversely affect the habitats and bioregions they invade economically, environmentally, and/or ecologically. Such invasive species may be either plants or animals and may drupt by dominating a region, wilderness areas, particular habitats, or wildland-urban interface land from loss of natural controls (such as predators or herbivores).

Chlorofluorocarbon	A chlorofluorocarbon is an organic compound that contains carbon, chlorine, and fluorine, produced as a volatile derivative of methane and ethane. A common subclass are the hydrochlorofluorocarbons (HCFCs), which contain hydrogen, as well. They are also commonly known by the DuPont trade name Freon.
Ozone	Ozone or trioxygen, is a triatomic molecule, consisting of three oxygen atoms. It is an allotrope of oxygen that is much less stable than the diatomic allotrope (O_2). Ozone in the lower atmosphere is an air pollutant with harmful effects on the respiratory systems of animals and will burn sensitive plants; however, the ozone layer in the upper atmosphere is beneficial, preventing damaging ultraviolet light from reaching the Earth's surface.
Ozone layer	The ozone layer is a layer in Earth's atmosphere which contains relatively high concentrations of ozone (O_3). This layer absorbs 97-99% of the Sun's high frequency ultraviet light, which potentially damages the life forms on Earth. It is mainly located in the lower portion of the stratosphere from approximately 20 to 30 kilometres (12 to 19 mi) above Earth, though the thickness varies seasonally and geographically.

Chapter 2. Environmental Systems: Matter, Energy, and Change

Catalyst	Catalysis is the change in rate of a chemical reaction due to the participation of a substance called a catalyst. Unlike other reagents that participate in the chemical reaction, a catalyst is not consumed by the reaction itself. A catalyst may participate in multiple chemical transformations.
Environmental mitigation	Environmental mitigation, compensatory mitigation, or mitigation banking, are terms used primarily by the United States government and the related environmental industry to describe projects or programs intended to offset known impacts to an existing historic or natural resource such as a stream, wetland, endangered species, archeological site or historic structure. To 'mitigate' means to make less harsh or hostile. Environmental mitigation is typically a part of an environmental crediting syst established by governing bodies which involves allocating debits and credits.
Montreal Protocol	The Montreal Protocol on Substances That Deplete the Ozone Layer (a protocol to the Vienna Convention for the Protection of the Ozone Layer) is an international treaty designed to protect the ozone layer by phasing out the production of numerous substances believed to be responsible for ozone depletion. The treaty was opened for signature on September 16, 1987, and entered into force on January 1, 1989, followed by a first meeting in Helsinki, May 1989. Since then, it has undergone seven revisions, in 1990 (London), 1991 (Nairobi), 1992 (Copenhagen), 1993 (Bangkok), 1995 (Vienna), 1997 (Montreal), and 1999 (Beijing). It is believed that if the international agreement is adhered to, the ozone layer is expected to recover by 2050. Due to its widespread adoption and implementation it has been hailed as an example of exceptional international co-operation, with Kofi Annan quoted as saying that 'perhaps the single most successful international agreement to date has been the Montreal Protocol'.

---

1. _____, radiant light and heat from the sun, has been harnessed by humans since ancient times using a range of ever-evolving technologies. Solar radiation, along with secondary solar-powered resources such as wind and wave power, hydroelectricity and biomass, account for most of the available renewable energy on earth. Only a minuscule fraction of the available _____ is used.

   a. Solar water heating
   b. Solar energy
   c. Gibbons v. Ogden
   d. Rhamnolipid

2. _____, is the ability of a liquid to flow against gravity where liquid spontaneously rises in a narrow space such as a thin tube, or in porous materials such as paper or in some non-porous materials such as liquified carbon fibre. This effect can cause liquids to flow against the force of gravity or the magnetic field induction. It occurs because of inter-molecular attractive forces between the liquid and solid surrounding surfaces; If the diameter of the tube is sufficiently small, then the combination of surface tension (which is caused by cohesion within the liquid) and forces of adhesion between the liquid and container act to lift the liquid.

   a. Cone of depression
   b. stream channel
   c. hadal zone
   d. Capillary action

3. In chemistry, _____ is the potential of a chemical substan to undergo a transformation through a chemical reaction or, to transform other chemical substans.Template:Fusion Breaking or making of chemical bonds involves energy, which may be either absorbed or evolved from a chemical system. Energy that can be released (or absorbed) because of a reaction between a set of chemical substans is equal to the differen between the energy content of the products and the reactants. This change in energy is change in internal energy of a chemical reaction.

   a. Chemical library
   b. Chemical energy
   c. Chemical similarity
   d. Chemical state

4. The _____ is a subatomic particle with the symbol p or p$^+$ and a positive electric charge of 1 elementary charge. One or more _____s are present in the nucleus of each atom, along with neutrons. The number of _____s in each atom is its atomic number.

---

a. Proton

b. Electron density

c. Electron diffraction

d. Electron excitation

5. The _____ is an expression of the tendency that over time, differences in temperature, pressure, and chemical potential equilibrate in an isolated physical system. From the state of thermodynamic equilibrium, the law deduced the principle of the increase of entropy and explains the phenomenon of irreversibility in nature. The second law declares the impossibility of machines that generate usable energy from the abundant internal energy of nature by processes called perpetual motion of the second kind.

a. Thermophoresis

b. Two-dimensional gas

c. Second law of thermodynamics

d. Chemical state

---

1. b
2. d
3. b
4. a
5. c

You can take the complete Chapter Practice Test

for Chapter 2. Environmental Systems: Matter, Energy, and Change
on all key terms, persons, places, and concepts.

Online 99 Cents

http://www.epub89.16.20190.2.cram101.com/

Use www.Cram101.com for all your study needs

including Cram101's online interactive problem solving labs in chemistry, statistics, mathematics, and more.

National park

Biodiversity

Biodiversity hotspot

Hotspot

Atmosphere

Autotroph

Cellular respiration

Natural resource

Photosynthesis

Convection

Energy flow

Isotope

Food chain

Food web

Heterotroph

Primary

Trophic level

Ethanol

Scavenger

_____ | Decomposer

_____ | Detritivore

_____ | Efficiency

_____ | Biosphere

_____ | Biogeochemical cycle

_____ | Carbon cycle

_____ | Evapotranspiration

_____ | Transpiration

_____ | Air pollution

_____ | Pollution

_____ | Carbon dioxide

_____ | Global change

_____ | Nitrogen cycle

_____ | Nitrogen fixation

_____ | Nutrient

_____ | Eutrophication

_____ | Algal bloom

_____ | Dead zone

_____ | Denitrification

_____ | Green Revolution _____

_____ | Phosphorus cycle _____

_____ | Sulfur _____

_____ | Sulfuric acid _____

_____ | Hurricane _____

_____ | Restoration ecology _____

_____ | Ecology _____

_____ | Hypothesis _____

_____ | Ecosystem services _____

_____ | Sustainability _____

National park	A national park is a reserve of natural, semi-natural, or developed land that a sovereign state declares or owns. Although individual nations designate their own national parks differently , an international organization, the International Union for Conservation of Nature (IUCN), and its World Commission on Protected Areas, has defined National Parks as its category II type of protected areas.
Biodiversity	Biodiversity is the degree of variation of life forms within a given species, ecosystem, biome, or an entire planet. Biodiversity is a measure of the health of ecosystems. Biodiversity is in part a function of climate.
Biodiversity hotspot	A biodiversity hotspot is a biogeographic region with a significant reservoir of biodiversity that is under threat from humans. The concept of biodiversity hotspots was originated by Norman Myers in two articles in 'The Environmentalist' (1988 ' 1990), revised after thorough analysis by Myers and others in 'Hotspots: Earth's Biologically Richest and Most Endangered Terrestrial Ecoregions'. To qualify as a biodiversity hotspot on Myers 2000 edition of the hotspot-map, a region must meet two strict criteria: it must contain at least 0.5% or 1,500 species of vascular plants as endemics, and it has to have lost at least 70% of its primary vegetation.
Hotspot	The places known as hotspots or hot spots in geology are volcanic regions thought to be fed by underlying mantle that is anomalously hot compared with the mantle elsewhere. They may be on, near to, or far from tectonic plate boundaries. There are two hypotheses to explain them.
Atmosphere	The standard atmosphere (symbol: atm) is an international reference pressure defined as 101325 Pa and formerly used as unit of pressure. For practical purposes it has been replaced by the bar which is 10^5 Pa. The difference of about 1% is not significant for many applications, and is within the error range of common pressure gges.
Autotroph	An autotroph, is an organism that produces complex organic compounds (such as carbohydrates, fats, and proteins) from simple inorganic molecules using energy from light (by photosynthesis) or inorganic chemical reactions (chemosynthesis). They are the producers in a food chain, such as plants on land or algae in water. They are able to make their own food and can fix carbon.

Cellular respiration	Cellular respiration is the set of the metabolic reactions and processes that take place in the cells of organisms to convert biochemical energy from nutrients into adenosine triphosphate (ATP), and then release waste products. The reactions involved in respiration are catabolic reactions that involve the redox reaction (oxidation of one molecule and the reduction of another). Respiration is one of the key ways a cell gains useful energy to fuel cellular changes.
Natural resource	Natural resources occur naturally within environments that exist relatively undisturbed by mankind, in a natural form. A natural resource is often characterized by amounts of biodiversity and geodiversity existent in various ecosystems. Natural resources are derived from the environment.
Photosynthesis	Photosynthesis is a chemical process that converts carbon dioxide into organic compounds, especially sugars, using the energy from sunlight. Photosynthesis occurs in plants, algae, and many species of bacteria, but not in archaea. Photosynthetic organisms are called photoautotrophs, since they can create their own food.
Convection	Convection is the concerted, collective movement of ensembles of molecules within fluids (i.e. liquids, gases) and rheids. Convection of mass cannot take place in solids, since neither bulk current flows nor significant diffusion can take place in solids. Diffusion of heat can take place in solids, but is referred to separately in that case as heat conduction.
Energy flow	In ecology, energy flow, also called the calorific flow, rrs to the flow of energy through a food chain. In an ecosystem, ecologists seek to quantify the relative importance of different component species and feeding relationships. A general energy flow scenario follows: <ul><li>Solar energy is fixed by the photoautotrophs, called primary producers, like green plants.</li></ul>
Isotope	Isotopes are variants of a particular chemical element. While all isotopes of a given element share the same number of protons, each isotope differs from the others in its number of neutrons. The term isotope is formed from the Greek roots isos (?σος 'equal') and topos (τ?πος 'place').

Chapter 3. Ecosystem Ecology: Interactions Between the Living and Nonliving World

Food chain	A food chain is somewhat a linear sequence of links in a food web starting from a trophic species that eats no other species in the web and ends at a trophic species that is eaten by no other species in the web. A food chain differs from a food web, because the complex polyphagous network of feeding relations are aggregated into trophic species and the chain only follows linear monophagous pathways. A common metric used to quantify food web trophic structure is food chain length.
Food web	A food web depicts feeding connections (what eats what) in an ecological community. Ecologists can broadly lump all life forms into one of two categories called trophic levels: 1) the autotrophs, and 2) the heterotrophs. To maintain their bodies, grow, develop, and to reproduce, autotrophs produce organic matter from inorganic substances, including both minerals and gases such as carbon dioxide.
Heterotroph	A heterotroph is an organism that cannot fix carbon and uses organic carbon for growth. This contrasts with autotrophs, such as plants and algae, which can use energy from sunlight (photoautotrophs) or inorganic compounds (lithoautotrophs) to produce organic compounds such as carbohydrates, fats, and proteins from inorganic carbon dioxide. These reduced carbon compounds can be used as an energy source by the autotroph and provide the energy in food consumed by heterotrophs.
Primary	A primary (or gravitational primary) is the main physical body of a gravitationally bound, multi-object system. This body contributes most of the mass of that system and will generally be located near its center of mass.
	In the solar system, the Sun is the primary for all objects that orbit around it.
Trophic level	The trophic level of an organism is the position it occupies in a food chain. A food chain represents a succession of organisms that eat another organism and are, in turn, eaten themselves. The number of steps an organism is from the start of the chain is a measure of its trophic level.
Ethanol	Ethanol, pure alcohol, grain alcohol, or drinking alcohol, is a volatile, flammable, colorless liquid. It is a psychoactive drug and one of the oldest recreational drugs. Best known as the type of alcohol found in alcoholic beverages, it is also used in thermometers, as a solvent, and as a fuel.

Scavenger	Scavenging is both a carnivorous and herbivorous feeding behaviour in which individual scavengers search out dead animal (corpses or carrion) and dead plant biomass on which to feed . The eating of carrion from the same species is referred to as cannibalism. Scavengers play an important role in the ecosystem by contributing to the decomposition of dead animal and plant material.
Decomposer	Decomposers (or saprotrophs) are organisms that break down dead or decaying organisms, and in doing so carry out the natural process of decomposition. Like herbivores and predators, decomposers are heterotrophic, meaning that they use organic substrates to get their energy, carbon and nutrients for growth and development. Decomposers use deceased organisms and non-living organic compounds as their food source.
Detritivore	Detritivores, also known as detritophages or detritus feeders or detritus eaters or saprophages, are heterotrophs that obtain nutrients by consuming detritus (decomposing organic matter). By doing so, they contribute to decomposition and the nutrient cycles. They should be distinguished from other decomposers, such as many species of bacteria, fungi and protists, unable to ingest discrete lumps of matter, instead live by absorbing and metabolising on a molecular scale.
Efficiency	Efficiency in general describes the extent to which time or effort is well used for the intended task or purpose. It is often used with the specific purpose of relaying the capability of a specific application of effort to produce a specific outcome effectively with a minimum amount or quantity of waste, expense, or unnecessary effort. 'Efficiency' has widely varying meanings in different disciplines.
Biosphere	The biosphere is the global sum of all ecosystems. It can also be called the zone of life on Earth, a closed (apart from solar and cosmic radiation) and self-regulating system. From the broadest biophysiological point of view, the biosphere is the global ecological system integrating all living beings and their relationships, including their interaction with the elements of the lithosphere, hydrosphere and atmosphere.
Biogeochemical cycle	In ecology and Earth science, a biogeochemical cycle is a pathway by which a chemical element or molecule moves through both biotic (biosphere) and abiotic (lithosphere, atmosphere, and hydrosphere) compartments of Earth. A cycle is a series of change which comes back to the starting point and which can be repeated.

The term 'biogeochemical' tells us that biological; geological and chemical factors are all involved.

Carbon cycle

The carbon cycle is the biogeochemical cycle by which carbon is exchanged among the biosphere, pedosphere, geosphere, hydrosphere, and atmosphere of the Earth. It is one of the most important cycles of the earth and allows for carbon to be recycled and reused throughout the biosphere and all of its organisms.

The carbon cycle was initially discovered by Joseph Priestley and Antoine Lavoisier, and popularized by Humphry Davy.

Evapotranspiration

Evapotranspiration is a term used to describe the sum of evaporation and plant transpiration from the Earth's land surface to atmosphere. Evaporation accounts for the movement of water to the air from sources such as the soil, canopy interception, and waterbodies. Transpiration accounts for the movement of water within a plant and the subsequent loss of water as vapor through stomata in its leaves.

Transpiration

Transpiration is a process similar to evaporation. It is a part of the water cycle, and it is the loss of water vapor from parts of plants (similar to sweating), especially in leaves but also in stems, flowers and roots. Leaf surfaces are dotted with openings which are collectively called stomata, and in most plants they are more numerous on the undersides of the foliage.

Air pollution

Air pollution is the introduction of chemicals, particulate matter, or biological materials that cause harm or discomfort to humans or other living organisms, or cause damage to the natural environment or built environment, into the atmosphere.

The atmosphere is a complex dynamic natural gaseous system that is essential to support life on planet Earth. Stratospheric ozone depletion due to air pollution has long been recognized as a threat to human health as well as to the Earth's ecosystems.

Pollution	Pollution is the introduction of contaminants into a natural environment that causes instability, disorder, harm or discomfort to the ecosystem i.e. physical systems or living organisms. Pollution can take the form of chemical substances or energy, such as noise, heat or light. Pollutants, the components of pollution, can be either foreign substances/energies or naturally occurring contaminants.
Carbon dioxide	Carbon dioxide is a naturally occurring chemical compound composed of two oxygen atoms covalently bonded to a single carbon atom. It is a gas at standard temperature and pressure and exists in Earth's atmosphere in this state, as a trace gas at a concentration of 0.039% by volume.
	As part of the carbon cycle known as photosynthesis, plants, algae, and cyanobacteria absorb carbon dioxide, light, and water to produce carbohydrate energy for themselves and oxygen as a waste product.
Global change	Global change refers to planetary-scale changes in the Earth system. The system consists of the land, oceans, atmosphere, poles, life, the planet's natural cycles and deep Earth processes. These constituent parts influence one another.
Nitrogen cycle	The nitrogen cycle is the process by which nitrogen is converted between its various chemical forms. This transformation can be carried out via both biological and non-biological processes. Important processes in the nitrogen cycle include fixation, mineralization, nitrification, and denitrification.
Nitrogen fixation	Nitrogen fixation is a process, biological, abiotic, or synthetic by which nitrogen (N_2) in the atmosphere is converted into ammonia (NH_3). Atmospheric nitrogen or elemental nitrogen (N_2) is relatively inert: it does not easily react with other chemicals to form new compounds. Fixation processes free up the nitrogen atoms from their diatomic form (N_2) to be used in other ways.
Nutrient	A nutrient is a chemical that an organism needs to live and grow or a substance used in an organism's metabolism which must be taken in from its environment. They are used to build and repair tissues, regulate body processes and are converted to and used as energy. Methods for nutrient intake vary, with animals and protists consuming foods that are digested by an internal digestive system, but most plants ingest nutrients directly from the soil through their roots or from the atmosphere.

Chapter 3. Ecosystem Ecology: Interactions Between the Living and Nonliving World

Eutrophication	Eutrophication, is the ecosystem response to the addition of artificial or natural substances, such as nitrates and phosphates, through fertilizers or sewage, to an aquatic system. One example is the 'bloom' or great increase of phytoplankton in a water body as a response to increased levels of nutrients. Negative environmental effects include hypoxia, the depletion of oxygen in the water, which induces reductions in specific fish and other animal populations.
Algal bloom	An algal bloom is a rapid increase or accumulation in the population of algae in an aquatic system. Algal blooms may occur in freshwater as well as marine environments. Typically, only one or a small number of phytoplankton species are involved, and some blooms may be recognized by discoloration of the water resulting from the high density of pigmented cells.
Dead zone	Dead zones are hypoxic (low-oxygen) areas in the world's oceans, the observed incidences of which have been increasing since oceanographers began noting them in the 1970s. These occur near inhabited coastlines, where aquatic life is most concentrated. (The vast middle portions of the oceans which naturally have little life are not considered 'dead zones'). The term can also be applied to the identical phenomenon in large lakes.
Denitrification	Denitrification is a microbially facilitated process of nitrate reduction that may ultimately produce molecular nitrogen (N_2) through a series of intermediate gaseous nitrogen oxide products. This respiratory process reduces oxidized forms of nitrogen in response to the oxidation of an electron donor such as organic matter. The preferred nitrogen electron acceptors in order of most to least thermodynamically favorable include nitrate (NO_3^-), nitrite (NO_2^-), nitric oxide (NO), and nitrous oxide (N_2O).
Green Revolution	Green Revolution refers to a series of research, development, and technology transfer initiatives, occurring between the 1940s and the late 1970s, that increased agriculture production around the world, beginning most markedly in the late 1960s. The initiatives, led by Norman Borlaug, the 'Father of the Green Revolution' credited with saving over a billion people from starvation, involved the development of high-yielding varieties of cereal grains, expansion of irrigation infrastructure, modernization of management techniques, distribution of hybridized seeds, synthetic fertilizers, and pesticides to farmers.

	The term 'Green Revolution' was first used in 1968 by former United States Agency for International Development (USAID) director William Gaud, who noted the spread of the new technologies and said, These and other developments in the field of agriculture contain the makings of a new revolution.
Phosphorus cycle	The phosphorus cycle is the biogeochemical cycle that describes the movement of phosphorus through the lithosphere, hydrosphere, and biosphere. Unlike many other biogeochemical cycles, the atmosphere does not play a significant role in the movement of phosphorus, because phosphorus and phosphorus-based compounds are usually solids at the typical ranges of temperature and pressure found on Earth. The production of phosphine gas occurs only in specialized, local conditions.
Sulfur	Sulfur or sulphur is the chemical element with atomic number 16. In the periodic table it is represented by the symbol S. It is an abundant, multivalent non-metal. Under normal conditions, sulfur atoms form cyclic octatomic molecules with chemical formula S_8. Elemental sulfur is a bright yellow crystalline solid when at room temperature.
Sulfuric acid	Sulfuric acid is a highly corrosive strong mineral acid with the molecular formula H_2SO_4. The historical name of this acid is oil of vitriol. It is a colorless to slightly yellow viscous liquid and is soluble in water at all concentrations.
Hurricane	Hurricane! (episode: 1616 (308)) is a Nova episode that aired on November 7, 1989 on PBS. The episode describes the fury of a hurricane and the history of hurricane forecasting. The episode features footage of Hurricane Camille of 1969 and Hurricane Gilbert of 1988 and behind the scenes footage at the National Hurricane Center as forecasters tracked Hurricane Gilbert from its formation to its landfall in northern Mexico. Notable meteorologists, Hugh Willoughby, Bob Sheets (then director of the National Hurricane Center) and Jeff Masters were shown in the episode.

Restoration ecology	Restoration ecology is the scientific study and practice of renewing and restoring degraded, damaged, or destroyed ecosystems and habitats in the environment by active human intervention and action. Restoration ecology emerged as a separate field in ecology in the 1980s. History Land managers, laypeople, and stewards have been practicing restoration for many hundreds, if not thousands of years, yet the scientific field of 'restoration ecology' was first identified and coined in the late 1980s by John Aber and William Jordan.
Ecology	Ecology is the scientific study of the relations that living organisms have with respect to each other and their natural environment. Variables of interest to ecologists include the composition, distribution, amount (biomass), number, and changing states of organisms within and among ecosystems. Ecosystems are hierarchical systems that are organized into a graded series of regularly interacting and semi-independent parts (e.g., species) that aggregate into higher orders of complex integrated wholes (e.g., communities).
Hypothesis	A hypothesis is a proposed explanation for a phenomenon. The term derives from the Greek, ? ποτιθ?ναι - hypotithenai meaning 'to put under' or 'to suppose'. For a hypothesis to be put forward as a scientific hypothesis, the scientific method requires that one can test it.
Ecosystem services	Humankind benefits from a multitude of resources and processes that are supplied by natural ecosystems. Collectively, these benefits are known as ecosystem services and include products like clean drinking water and processes such as the decomposition of wastes. While scientists and environmentalists have discussed ecosystem services for decades, these services were popularized and their definitions formalized by the United Nations 2004 Millennium Ecosystem Assessment (MA), a four-year study involving more than 1,300 scientists worldwide.
Sustainability	Sustainability is the capacity to endure. For humans, sustainability is the long-term maintenance of responsibility, which has environmental, economic, and social dimensions, and encompasses the concept of stewardship, the responsible management of resource use. In ecology, sustainability describes how biological systems remain diverse and productive over time, a necessary precondition for human well-being.

---

1. An _____ is a rapid increase or accumulation in the population of algae in an aquatic system. _____s may occur in freshwater as well as marine environments. Typically, only one or a small number of phytoplankton species are involved, and some blooms may be recognized by discoloration of the water resulting from the high density of pigmented cells.

   a. Algal bloom
   b. Index of biological integrity
   c. Odyssey
   d. Industrial wastewater treatment

2. Scavenging is both a carnivorous and herbivorous feeding behaviour in which individual _____s search out dead animal (corpses or carrion) and dead plant biomass on which to feed . The eating of carrion from the same species is referred to as cannibalism. _____s play an important role in the ecosystem by contributing to the decomposition of dead animal and plant material.

   a. Vulture
   b. Uranium
   c. Scavenger
   d. tropical climate

3. _____s (or saprotrophs) are organisms that break down dead or decaying organisms, and in doing so carry out the natural process of decomposition. Like herbivores and predators, _____s are heterotrophic, meaning that they use organic substrates to get their energy, carbon and nutrients for growth and development. _____s use deceased organisms and non-living organic compounds as their food source.

   a. Decomposer
   b. Biodiversity hotspot
   c. Biodiversity Indicators Partnership
   d. Bonn Convention

4. The places known as _____s or hot spots in geology are volcanic regions thought to be fed by underlying mantle that is anomalously hot compared with the mantle elsewhere. They may be on, near to, or far from tectonic plate boundaries. There are two hypotheses to explain them.

   a. Juglone
   b. Bonn Convention
   c. Breeding season
   d. Hotspot

---

5. _____ is the scientific study of the relations that living organisms have with respect to each other and their natural environment. Variables of interest to ecologists include the composition, distribution, amount (biomass), number, and changing states of organisms within and among ecosystems. Ecosystems are hierarchical systems that are organized into a graded series of regularly interacting and semi-independent parts (e.g., species) that aggregate into higher orders of complex integrated wholes (e.g., communities).

    a. Ecologist
    b. Eidonomy
    c. Ecology
    d. Engorgement

---

1. a
2. c
3. a
4. d
5. c

You can take the complete Chapter Practice Test

for Chapter 3. Ecosystem Ecology: Interactions Between the Living and Nonliving World
on all key terms, persons, places, and concepts.

Online 99 Cents

http://www.epub89.16.20190.3.cram101.com/

Use www.Cram101.com for all your study needs

including Cram101's online interactive problem solving labs in chemistry, statistics, mathematics, and more.

CHAPTER OUTLINE: KEY TERMS, PEOPLE, PLACES, CONCEPTS
Chapter 4
Global Climates and Biomes: Geographic Variations in Temperature

55

Drought

Dust Bowl

Atmosphere

Exosphere

Mesosphere

Ozone

Ozone layer

Stratosphere

Thermosphere

Troposphere

Convection

Radiation

Carrying capacity

Albedo

Landfill

Coriolis

Coriolis effect

Grassland

Hadley cell

Chapter 4. Global Climates and Biomes: Geographic Variations in Temperature

_____ | Biome ____

_____ | Convergence _____

_____ | Convergence zone ____

_____ | Prevailing winds ____

_____ | Rotation ____

_____ | Trade wind ____

_____ | Solstice ____

_____ | Tropic of Cancer ____

_____ | Tropic of Capricorn ____

_____ | Westerlies ____

_____ | Energy current ____

_____ | Thermohaline circulation ____

_____ | Upwelling ____

_____ | Global warming ____

_____ | Gulf Stream ____

_____ | Rain shadow ____

_____ | Climate change ____

_____ | Global change ____

_____ | Growing season ____

	Permafrost
	Tundra
	Temperate rainforest
	Tropical rainforest
	Benthic zone
	Limnetic zone
	Littoral zone
	Phytoplankton
	Profundal zone
	Wetland
	Ecosystem services
	Intertidal zone
	Coral bleaching
	Coral reef
	Invertebrate
	Chemosynthesis
	Photic zone
	Sustainability
	Shade-grown coffee

Drought	A drought is an extended period of months or years when a region notes a deficiency in its water supply whether surface or underground water. Generally, this occurs when a region receives consistently below average precipitation. It can have a substantial impact on the ecosystem and agriculture of the affected region.
Dust Bowl	The Dust Bowl, was a period of severe dust storms causing major ecological and agricultural damage to American and Canadian prairie lands from 1930 to 1936 (in some areas until 1940). The phenomenon was caused by severe drought coupled with decades of extensive farming without crop rotation, fallow fields, cover crops or other techniques to prevent wind erosion. Deep plowing of the virgin topsoil of the Great Plains had displaced the natural deep-rooted grasses that normally kept the soil in place and trapped moisture even during periods of drought and high winds.
Atmosphere	The standard atmosphere (symbol: atm) is an international reference pressure defined as 101325 Pa and formerly used as unit of pressure. For practical purposes it has been replaced by the bar which is 10^5 Pa. The difference of about 1% is not significant for many applications, and is within the error range of common pressure gges.
Exosphere	The exosphere is the uppermost layer of the atmosphere. In the exosphere, an upward travelling molecule moving fast enough to attain escape velocity can escape to space with a low chance of collisions; if it is moving below escape velocity it will be prevented from escaping from the celestial body by gravity. In either case, such a molecule is unlikely to collide with another molecule due to the exosphere's low density.
Mesosphere	The mesosphere refers to the mantle in the region between the asthenosphere and the outer core. The upper boundary is defined as the sharp increase in seismic wave velocities and density at a depth of 660 km. As depth increases, pressure builds and forces atoms into a denser, more rigid structure; thus the difference between mesosphere and asthenosphere is likely due to density and rigidity differences, that is, physical factors, and not to any difference in chemical composition.
Ozone	Ozone or trioxygen, is a triatomic molecule, consisting of three oxygen atoms. It is an allotrope of oxygen that is much less stable than the diatomic allotrope (O_2). Ozone in the lower atmosphere is an air pollutant with harmful effects on the respiratory systems of animals and will burn sensitive plants; however, the ozone layer in the upper atmosphere is beneficial, preventing damaging ultraviolet light from reaching the Earth's surface.

Ozone layer	The ozone layer is a layer in Earth's atmosphere which contains relatively high concentrations of ozone (O_3). This layer absorbs 97-99% of the Sun's high frequency ultraviet light, which potentially damages the life forms on Earth. It is mainly located in the lower portion of the stratosphere from approximately 20 to 30 kilometres (12 to 19 mi) above Earth, though the thickness varies seasonally and geographically.
Stratosphere	The stratosphere is the second major layer of Earth's atmosphere, just above the troposphere, and below the mesosphere. It is stratified in temperature, with warmer layers higher up and cooler layers farther down. This is in contrast to the troposphere near the Earth's surface, which is cooler higher up and warmer farther down.
Thermosphere	The thermosphere is the layer of the Earth's atmosphere directly above the mesosphere and directly below the exosphere. Within this layer, ultraviolet radiation causes ionization. The International Space Station has a stable orbit within the middle of the thermosphere, between 320 and 380 kilometres (200 and 240 mi).
Troposphere	The troposphere is the lowest portion of Earth's atmosphere. It contains approximately 75% of the atmosphere's mass and 99% of its water vapor and aerosols. The average depth of the troposphere is approximately 17 km (11 mi) in the middle latitudes.
Convection	Convection is the concerted, collective movement of ensembles of molecules within fluids (i.e. liquids, gases) and rheids. Convection of mass cannot take place in solids, since neither bulk current flows nor significant diffusion can take place in solids. Diffusion of heat can take place in solids, but is referred to separately in that case as heat conduction.
Radiation	In physics, radiation is a process in which energetic particles or energetic waves travel through a medium or space. Two types of radiation are commonly differentiated in the way they interact with normal chemical matter: ionizing and non-ionizing radiation. The word radiation is often colloquially used in reference to ionizing radiation but the term radiation may correctly also refer to non-ionizing radiation.

Chapter 4. Global Climates and Biomes: Geographic Variations in Temperature

Carrying capacity	The carrying capacity of a biological species in an environment is the maximum population size of the species that the environment can sustain indefinitely, given the food, habitat, water and other necessities available in the environment. In population biology, carrying capacity is defined as the environment's maximal load, which is different from the concept of population equilibrium. For the human population, more complex variables such as sanitation and medical care are sometimes considered as part of the necessary establishment.
Albedo	Albedo or reflection coefficient, derived from Latin albedo 'whiteness' (or reflected sunlight), in turn from albus 'white', is the diffuse reflectivity or reflecting power of a surface. It is defined as the ratio of reflected radiation from the surface to incident radiation upon it. Being a dimensionless fraction, it may also be expressed as a percentage, and is measured on a scale from zero for no reflecting power of a perfectly black surface, to 1 for perfect reflection of a white surface.
Landfill	A landfill site (also known as tip, dump or rubbish dump and historically as a midden) is a site for the disposal of waste materials by burial and is the oldest form of waste treatment. Historically, landfills have been the most common methods of organized waste disposal and remain so in many places around the world. Landfills may include internal waste disposal sites (where a producer of waste carries out their own waste disposal at the place of production) as well as sites used by many producers.
Coriolis	The Coriolis satellite is a Naval Research Laboratory and Air Force Research Laboratory earth and space observation satellite launched from Vandenberg Air Force Base, on 2003-01-06 at 14:19 GMT. Instruments Windsat WINDSAT is a joint Integrated Program OfficeDepartment of Defense demonstration project, intended to measure ocean surface wind speed and wind direction from space using a polarimetric radiometer. Solar Mass Ejection Imager (SMEI)

The Solar Mass Ejection Imager (SMEI) is an instrument intended to detect disturbances in the solar wind by means of imaging scattered light from the free electrons in the plasma of the solar wind. To do this three CCD cameras observe sections of the sky of size 60 by 3 degree.

Coriolis effect

In physics, the Coriolis effect is a deflection of moving objects when they are viewed in a rotating referen frame. In a referen frame with clockwise rotation, the deflection is to the left of the motion of the object; in one with counter-clockwise rotation, the deflection is to the right. The mathematical expression for the Coriolis for appeared in an 1835 paper by French scientist Gaspard-Gustave Coriolis, in connection with the theory of water wheels, and also in the tidal equations of Pierre-Simon Lapla in 1778. And even earlier, Italian scientists Giovanni Battista Riccioli and his assistant Fransco Maria Grimaldi described the effect in connection with artillery in the 1651 Almagestum Novum, writing that rotation of the Earth should cause a cannon ball fired to the north to deflect to the east.

Grassland

Grasslands are areas where the vegetation is dominated by grasses (Poaceae) and other herbaceous (non-woody) plants (forbs). However, sedge (Cyperaceae) and rush (Juncaceae) families can also be found. Grasslands occur naturally on all continents except Antarctica. In temperate latitudes, such as northwestern Europe and the Great Plains and California in North America, native grasslands are dominated by perennial bunch grass species, whereas in warmer climates annual species form a greater component of the vegetation.

Hadley cell

The Hadley cell, is a tropical atmospheric circulation that is defined by the average over longitude, which features rising motion near the equator, poleward flow 10-15 kilometers above the surface, descending motion in the subtropics, and equatorward flow near the surface. This circulation is intimately related to the trade winds, tropical rainbelts, subtropical deserts and the jet streams.

There are three primary circulation cells.

Biome

Biomes are climatically and geographically defined as similar climatic conditions on the Earth, such as communities of plants, animals, and soil organisms, and are often referred to as ecosystems. Some parts of the earth have more or less the same kind of abiotic and biotic factors spread over a large area, creating a typical ecosystem over that area. Such major ecosystems are termed as biomes.

Convergence	Convergence in sustainability sciences refers to mechanisms and pathways that lead towards sustainability with a specific focus on 'Equity within biological planetary limits'. These pathways and mechanisms explicitly advocate equity and recognise the need for redistribution of the Earth's resources in order for human society to operate enduringly within the Earth's biophysical limits.
	This use of the term 'convergence' harks from the concept of contraction and convergence (C'C), taking its core principles of Equity and Survival and applying them beyond the frame of greenhouse gas emissions to the wider sustainability agenda.
Convergence zone	Convergence zone usually refers to a region in the atmosphere where two prevailing flows meet and interact, usually resulting in distinctive weather conditions.
	An example of a convergence zone is the Intertropical Convergence Zone(ITCZ), a low pressure area which girdles the Earth at the Equator. Another example is the South Pacific convergence zone that extends from the western Pacific Ocean toward French Polynesia.
Prevailing winds	Prevailing winds are winds that blow predominantly from a single general direction over a particular point on Earth's surface. The dominant winds are the trends in direction of wind with the highest speed over a particular point on the Earth's surface. A region's prevailing and dominant winds are often affected by global patterns of movement in the Earth's atmosphere.
Rotation	A rotation is a circular movement of an object around a center (or point) of rotation. A three-dimensional object rotates always around an imaginary line called a rotation axis. If the axis is within the body, and passes through its center of mass the body is said to rotate upon itself, or spin.

Trade wind	The trade winds (also called trades) are the prevailing pattern of easterly surface winds found in the tropics, within the lower portion of the Earth's atmosphere, in the lower section of the troposphere near the Earth's equator. The trade winds blow predominantly from the northeast in the Northern Hemisphere and from the southeast in the Southern Hemisphere, strengthening during the winter and when the Arctic oscillation is in its warm phase. Historically, the trade winds have been used by captains of sailing ships to cross the world's oceans for centuries, and enabled European empire expansion into the Americas and trade routes to become established across the Atlantic and Pacific oceans.
Solstice	A solstice is an astronomical event that happens twice each year when the Sun reaches its highest position in the sky as seen from the North or South Pole. The word solstice is derived from the Latin sol (sun) and sistere (to stand still), because at the solstices, the Sun stands still in declination; that is, the seasonal movement of the Sun's path comes to a stop before reversing direction. The solstices, together with the equinoxes, are connected with the seasons.
Tropic of Cancer	The Tropic of Cancer, also referred to as the Northern tropic, is the circle of latitude on the Earth that marks the most northerly position at which the Sun may appear directly overhead at its zenith. This event occurs once per year, at the time of the June solstice, when the Northern Hemisphere is tilted toward the Sun to its maximum extent. Its Southern Hemisphere counterpart, marking the most southerly position at which the Sun may appear directly overhead, is the Tropic of Capricorn.
Tropic of Capricorn	The Tropic of Capricorn, marks the most southerly latitude at which the Sun can appear directly overhead. This event occurs at the December solstice, when the southern hemisphere is tilted towards the Sun to its maximum extent. Tropic of Capricorn is one of the five major circles of latitude that mark maps of the Earth.

Westerlies	The Westerlies, anti-trades, or Prevailing Westerlies, are the prevailing winds in the middle latitudes between 30 and 60 degrees latitude, blowing from the high pressure area in the horse latitudes towards the poles. These prevailing winds blow from the west to the east, and steer extratropical cyclones in this general manner. Tropical cyclones which cross the subtropical ridge axis into the Westerlies recurve due to the increased westerly flow.
Energy current	Energy current is a flow of energy defined by the Poynting vtor (E×H), as opposed to normal current (flow of charge). It was originally postulated by Oliver Heaviside. Explanation 'Energy current' is a somewhat informal term that is used, on occasion, to describe the process of energy transfer in situations where the transfer can usefully be viewed in terms of a flow.
Thermohaline circulation	The term thermohaline circulation refers to the part of the large-scale ocean circulation that is driven by global density gradients created by surface heat and freshwater fluxes. Wind-driven surface currents (such as the Gulf Stream) head polewards from the equatorial Atlantic Ocean, cooling all the while and eventually sinking at high latitudes (forming North Atlantic Deep Water). This dense water then flows into the ocean basins.
Upwelling	Upwelling is an oceanographic phenomenon that involves wind-driven motion of dense, cooler, and usually nutrient-rich water towards the ocean surface, replacing the warmer, usually nutrient-depleted surface water. The increased availability in upwelling regions results in high levels of primary productivity and thus fishery production. Approximately 25% of the total global marine fish catches come from five upwellings that occupy only 5% of the total ocean area.
Global warming	Global warming refers to the rising average temperature of Earth's atmosphere and oceans, which began to increase in the late 19th century and is projected to continue rising. Since the early 20th century, Earth's average surface temperature has increased by about 0.8 °C (1.4 °F), with about two thirds of the increase occurring since 1980. Warming of the climate system is unequivocal, and scientists are more than 90% certain that most of it is caused by increasing concentrations of greenhouse gases produced by human activities such as deforestation and the burning of fossil fuels. These findings are recognized by the national science academies of all major industrialized nations.[A]

Climate model projections are summarized in the 2007 Fourth Assessment Report (AR4) by the Intergovernmental Panel on Climate Change (IPCC).

Gulf Stream	The Gulf Stream, together with its northern extension towards Europe, the North Atlantic Drift, is a powerful, warm, and swift Atlantic ocean current that originates at the tip of Florida, and follows the eastern coastlines of the United States and Newfoundland before crossing the Atlantic Ocean. The process of western intensification causes the Gulf Stream to be a northward accelerating current off the east coast of North America. At about , it splits in two, with the northern stream crossing to northern Europe and the southern stream recirculating off West Africa. The Gulf Stream influences the climate of the east coast of North America from Florida to Newfoundland, and the west coast of Europe. Although there has been recent debate, there is consensus that the climate of Western Europe and Northern Europe is warmer than it would otherwise be due to the North Atlantic drift, one of the branches from the tail of the Gulf Stream.
Rain shadow	A rain shadow is a dry area on the lee side of a mountainous area. The mountains block the passage of rain-producing weather systems, casting a 'shadow' of dryness behind them. As shown by the diagram to the right, the warm moist air is 'pulled' by the prevailing winds over a mountain.
Climate change	Climate change is a significant and lasting change in the statistical distribution of weather patterns over periods ranging from decades to millions of years. It may be a change in average weather conditions or the distribution of events around that average (e.g., more or fewer extreme weather events). Climate change may be limited to a specific region or may our across the whole Earth.
Global change	Global change refers to planetary-scale changes in the Earth system. The system consists of the land, oceans, atmosphere, poles, life, the planet's natural cycles and deep Earth processes. These constituent parts influence one another.
Growing season	In botany, horticulture, and agriculture the growing season is the period of each year when native plants and ornamental plants grow; and when crops can be grown. The growing season is usually determined by climate and elevation, and in horticulture and agriculture the plant-crop selection. Depending on the location, temperature, daylight hours (photoperiod), and rainfall, may all be critical environmental factors.

Permafrost	In geology, permafrost is soil at or below the freezing point of water 0 °C (32 °F) for two or more years. Ice is not always present, as may be in the case of nonporous bedrock, but it frequently occurs and it may be in amounts exceeding the potential hydraulic saturation of the ground material. Most permafrost is located in high latitudes (i.e. land close to the North and South poles), but alpine permafrost may exist at high altitudes in much lower latitudes.
Tundra	In physical geography, tundra is a biome where the tree growth is hindered by low temperatures and short growing seasons. The term tundra comes through Russian тундра from the Kildin Sami word tundâr 'uplands,' 'treeless mountain tract.' There are three types of tundra: Arctic tundra, alpine tundra, and Antarctic tundra. In tundra, the vegetation is composed of dwarf shrubs, sedges and grasses, mosses, and lichens.
Temperate rainforest	Temperate rainforests are coniferous or broadleaf forests that occur in the temperate zone and receive high rainfall.

For temperate rain forests of North America, Alaback's definition is widely recognized:

1. Annual precipitation 200-400 cm
2. Mean annual temperature between $4^{o}C$ and $12^{o}C$. (39° and 54° Fahrenheit)

However, required annual precipitation depends on factors such as distribution of rainfall over the year, temperatures over the year and fog presence, and definitions in other countries differ considerably. For example, Australian definitions are ecological-structural rather than climatic:

1. Closed canopy of trees excludes at least 70% of the sky
2. Forest is composed mainly of tree species which do not require fire for regeneration, but with seedlings able to regenerate under shade and in natural openings

The latter would, for example, exclude a part of the temperate rain forests of western North America, as Coast Douglas-fir, one of its dominant tree species, requires stand-destroying disturbance to initiate a new cohort of seedlings.

Tropical rainforest	A tropical rainforest is a place roughly within 28 degrees north or south of the equator. They are found in Asia, Australia, Africa, South America, Central America, Mexico and on many of the Pacific Islands. Within the World Wildlife Fund's biome classification, tropical rainforests are thought to be a type of tropical wet forest (or tropical moist broadleaf forest) and may also be referred to as lowland equatorial evergreen rainforest.
Benthic zone	The benthic zone is the ecological region at the lowest level of a body of water such as an ocean or a lake, including the sediment surface and some sub-surface layers. Organisms living in this zone are called benthos. They generally live in close relationship with the substrate bottom; many such organisms are permanently attached to the bottom.
Limnetic zone	The limnetic zone is the well-lit, open surface waters in a lake, away from the shore. The vegetation of the littoral zone surrounds this expanse of open water and it is above the profundal zone. It can be defined as the lighted surface waters in the area where the lake bottom is too deep and unlit to support rooted aquatic plants.
Littoral zone	The littoral zone is that part of a sea, lake or river that is close to the shore. In coastal environments the littoral zone extends from the high water mark, which is rarely inundated, to shoreline areas that are permanently submerged. It always includes this intertidal zone and is often used to mean the same as the intertidal zone.
Phytoplankton	Phytoplankton are the autotrophic component of the plankton community. Most phytoplankton are too small to be individually seen with the unaided eye. However, when present in high enough numbers, they may appear as a green discoloration of the water due to the presence of chlorophyll within their cells (although the actual color may vary with the species of phytoplankton present due to varying levels of chlorophyll or the presence of accessory pigments such as phycobiliproteins, xanthophylls, etc.).
Profundal zone	The profundal zone is a very cold and ordinary zone, such as an ocean or a lake, located below the range of effective light penetration. This is typically below the thermocline, the vertical zone in the water through which temperature drops rapidly. The lack of light in the profundal zone determines the type of biological community that can live in this region, which is distinctly different from the community in the overlying waters.

Wetland	A wetland is a land area that is saturated with water, either permanently or seasonally, such that it takes on characteristics that distinguish it as a distinct ecosystem. The primary factor that distinguishes wetlands is the characteristic vegetation that is adapted to its unique soil conditions: Wetlands are made up primarily of hydric soil, which supports aquatic plants. The water found in wetlands can be saltwater, freshwater, or brackish.
Ecosystem services	Humankind benefits from a multitude of resources and processes that are supplied by natural ecosystems. Collectively, these benefits are known as ecosystem services and include products like clean drinking water and processes such as the decomposition of wastes. While scientists and environmentalists have discussed ecosystem services for decades, these services were popularized and their definitions formalized by the United Nations 2004 Millennium Ecosystem Assessment (MA), a four-year study involving more than 1,300 scientists worldwide.
Intertidal zone	The intertidal zone is the area that is above water at low tide and under water at high tide (for example, the area between tide marks). This area can include many different types of habitats, with many types of animals like starfish, sea urchins, and some species of coral. The well known area also includes steep rocky cliffs, sandy beaches, or wetlands (e.g., vast mudflats).
Coral bleaching	Coral bleaching is the loss of intracellular endosymbionts (Symbiodinium, also known as zooxanthellae) through either expulsion or loss of algal pigmentation. The corals that form the structure of the great reef ecosystems of tropical seas depend upon a symbiotic relationship with unicellular flagellate protozoa that are photosynthetic and live within their tissues. Zooxanthellae give coral its coloration, with the specific color depending on the particular clade.
Coral reef	Coral reefs are underwater structures made from calcium carbonate secreted by corals. Corals are colonies of tiny living animals found in marine waters that contain few nutrients. Most coral reefs are built from stony corals, which in turn consist of polyps that cluster in groups.
Invertebrate	An invertebrate is an animal without a backbone. The group includes 97% of all animal species - all animals except those in the chordate subphylum Vertebrata (fish, amphibians, reptiles, birds, and mammals).

Invertebrates form a paraphyletic group.

Chemosynthesis	In biochemistry, chemosynthesis is the biological conversion of one or more carbon molecules (usually carbon dioxide or methane) and nutrients into organic matter using the oxidation of inorganic molecules (e.g. hydrogen gas, hydrogen sulfide) or methane as a source of energy, rather than sunlight, as in photosynthesis. Chemoautotrophs, organisms that obtain carbon through chemosynthesis, are phylogenetically diverse, but groups that include conspicuous or biogeochemically-important taxa include the sulfur-oxidizing gamma and epsilon proteobacteria, the Aquificaeles, the Methanogenic archaea and the neutrophilic iron-oxidizing bacteria. Many microorganisms in dark regions of the oceans also use chemosynthesis to produce biomass from single carbon molecules.
Photic zone	The photic zone is exposed to sufficient sunlight for photosynthesis to occur. The depth of the photic zone can be affected greatly by seasonal turbidity. It extends from the atmosphere-water interface downwards to a depth where light intensity falls to one percent of that at the surface, called the euphotic depth.
Sustainability	Sustainability is the capacity to endure. For humans, sustainability is the long-term maintenance of responsibility, which has environmental, economic, and social dimensions, and encompasses the concept of stewardship, the responsible management of resource use. In ecology, sustainability describes how biological systems remain diverse and productive over time, a necessary precondition for human well-being.

| Shade-grown coffee | Shade-grown coffee is a form of the beverage produced from coffee plants grown under a canopy of trees. Because it incorporates principles of natural ecology to promote natural ecological relationships, shade-grown coffee can be considered an offshoot of agricultural permaculture.

History

Most of the original coffee trees brought to the New World from European countries would burn in the sun, which made shade necessary for growth. |

---

1. The _____ is a layer in Earth's atmosphere which contains relatively high concentrations of ozone (O_3). This layer absorbs 97-99% of the Sun's high frequency ultraviet light, which potentially damages the life forms on Earth. It is mainly located in the lower portion of the stratosphere from approximately 20 to 30 kilometres (12 to 19 mi) above Earth, though the thickness varies seasonally and geographically.

    a. Ozone Mapping and Profiler Suite
    b. Ozone layer
    c. Ozone Mapping and Profiler Suite
    d. Ozone-oxygen cycle

2. A _____ is an extended period of months or years when a region notes a deficiency in its water supply whether surface or underground water. Generally, this occurs when a region receives consistently below average precipitation. It can have a substantial impact on the ecosystem and agriculture of the affected region.

    a. Flood Control Act of 1928
    b. Flood-meadow
    c. Floodplain restoration
    d. Drought

3. The _____ refers to the mantle in the region between the asthenosphere and the outer core. The upper boundary is defined as the sharp increase in seismic wave velocities and density at a depth of 660 km. As depth increases, pressure builds and forces atoms into a denser, more rigid structure; thus the difference between _____ and asthenosphere is likely due to density and rigidity differences, that is, physical factors, and not to any difference in chemical composition.

    a. Seismogenic layer
    b. Mesosphere
    c. British Geophysical Association
    d. Canadian Geophysical Union

4. An _____ is an animal without a backbone. The group includes 97% of all animal species - all animals except those in the chordate subphylum Vertebrata (fish, amphibians, reptiles, birds, and mammals).

    _____s form a paraphyletic group.

---

a. Ischnochitonidae
b. Ivoechiton
c. Ocellochiton
d. Invertebrate

5. The _____, was a period of severe dust storms causing major ecological and agricultural damage to American and Canadian prairie lands from 1930 to 1936 (in some areas until 1940). The phenomenon was caused by severe drought coupled with decades of extensive farming without crop rotation, fallow fields, cover crops or other techniques to prevent wind erosion. Deep plowing of the virgin topsoil of the Great Plains had displaced the natural deep-rooted grasses that normally kept the soil in place and trapped moisture even during periods of drought and high winds.

a. Juglone
b. Flood-meadow
c. Floodplain restoration
d. Dust Bowl

---

1. b
2. d
3. b
4. d
5. d

You can take the complete Chapter Practice Test

for Chapter 4. Global Climates and Biomes: Geographic Variations in Temperature
on all key terms, persons, places, and concepts.

Online 99 Cents

http://www.epub89.16.20190.4.cram101.com/

Use www.Cram101.com for all your study needs

including Cram101's online interactive problem solving labs in chemistry, statistics, mathematics, and more.

Evolution and Biodiversity: Origin and Diversification of Organisms

Biodiversity

Species diversity

Species evenness

Species richness

Phylogenetic tree

Air pollution

Pollution

Speciation

Radiation

Green Revolution

Biodiversity hotspot

Hotspot

Domestication

Natural selection

Pesticide

Allopatric speciation

Commensalism

Extinction

Sympatric speciation

Chapter 5. Evolution and Biodiversity: Origin and Diversification of Organisms

_____ Environmental change

_____ Generation time

_____ Population size

_____ Genetically modified organism

_____ Invertebrate

_____ Species distribution

_____ Global change

_____ Fossil

_____ Sustainability

_____ Rights

_____ Habitat

_____ Threatened species

_____ National park

CHAPTER HIGHLIGHTS: KEY TERMS, PEOPLE, PLACES, CONCEPTS
Chapter 5. Evolution and Biodiversity: Origin and Diversification of Organisms

77

Biodiversity	Biodiversity is the degree of variation of life forms within a given species, ecosystem, biome, or an entire planet. Biodiversity is a measure of the health of ecosystems. Biodiversity is in part a function of climate.
Species diversity	Species diversity is the effective number of different species that are represented in a collection of individuals (a dataset). The effective number of species refers to the number of equally-abundant species needed to obtain the same mean proportional species abundance as that observed in the dataset of interest (where all species may not be equally abundant). Species diversity consists of two components, species richness and species evenness.
Species evenness	Species evenness refers to how close in numbers each species in an environment are. Mathematically it is defined as a diversity index, a measure of biodiversity which quantifies how equal the community is numerically. So if there are 40 foxes, and 1000 dogs, the community is not very even.
Species richness	Species richness is the number of different species in a given area. It is represented in equation form as S. Species richness is the fundamental unit in which to assess the homogeneity of an environment. Typically, species richness is used in conservation studies to determine the sensitivity of ecosystems and their resident species. The actual number of species calculated alone is largely an arbitrary number. These studies, therefore, often develop a rubric or measure for valuing the species richness number(s) or adopt one from previous studies on similar ecosystems.
Phylogenetic tree	A phylogenetic tree is a branching diagram or 'tree' showing the inferred evolutionary relationships among various biological species or other entities based upon similarities and differences in their physical and/or genetic characteristics. The taxa joined together in the tree are implied to have descended from a common ancestor. In a rooted phylogenetic tree, each node with descendants represents the inferred most recent common ancestor of the descendants, and the edge lengths in some trees may be interpreted as time estimates.
Air pollution	Air pollution is the introduction of chemicals, particulate matter, or biological materials that cause harm or discomfort to humans or other living organisms, or cause damage to the natural environment or built environment, into the atmosphere.

The atmosphere is a complex dynamic natural gaseous system that is essential to support life on planet Earth. Stratospheric ozone depletion due to air pollution has long been recognized as a threat to human health as well as to the Earth's ecosystems.

Pollution	Pollution is the introduction of contaminants into a natural environment that causes instability, disorder, harm or discomfort to the ecosystem i.e. physical systems or living organisms. Pollution can take the form of chemical substances or energy, such as noise, heat or light. Pollutants, the components of pollution, can be either foreign substances/energies or naturally occurring contaminants.
Speciation	Speciation is the evolutionary process by which new biological species arise. The biologist Orator F. Cook seems to have been the first to coin the term 'speciation' for the splitting of lineages or 'cladogenesis,' as opposed to 'anagenesis' or 'phyletic evolution' occurring within lineages. Whether genetic drift is a minor or major contributor to speciation is the subject matter of much ongoing discussion.
Radiation	In physics, radiation is a process in which energetic particles or energetic waves travel through a medium or space. Two types of radiation are commonly differentiated in the way they interact with normal chemical matter: ionizing and non-ionizing radiation. The word radiation is often colloquially used in reference to ionizing radiation but the term radiation may correctly also refer to non-ionizing radiation.
Green Revolution	Green Revolution refers to a series of research, development, and technology transfer initiatives, occurring between the 1940s and the late 1970s, that increased agriculture production around the world, beginning most markedly in the late 1960s.
	The initiatives, led by Norman Borlaug, the 'Father of the Green Revolution' credited with saving over a billion people from starvation, involved the development of high-yielding varieties of cereal grains, expansion of irrigation infrastructure, modernization of management techniques, distribution of hybridized seeds, synthetic fertilizers, and pesticides to farmers.
	The term 'Green Revolution' was first used in 1968 by former United States Agency for International Development (USAID) director William Gaud, who noted the spread of the new technologies and said,

---

	These and other developments in the field of agriculture contain the makings of a new revolution.
Biodiversity hotspot	A biodiversity hotspot is a biogeographic region with a significant reservoir of biodiversity that is under threat from humans.
	The concept of biodiversity hotspots was originated by Norman Myers in two articles in 'The Environmentalist' (1988 ' 1990), revised after thorough analysis by Myers and others in 'Hotspots: Earth's Biologically Richest and Most Endangered Terrestrial Ecoregions'.
	To qualify as a biodiversity hotspot on Myers 2000 edition of the hotspot-map, a region must meet two strict criteria: it must contain at least 0.5% or 1,500 species of vascular plants as endemics, and it has to have lost at least 70% of its primary vegetation.
Hotspot	The places known as hotspots or hot spots in geology are volcanic regions thought to be fed by underlying mantle that is anomalously hot compared with the mantle elsewhere. They may be on, near to, or far from tectonic plate boundaries. There are two hypotheses to explain them.
Domestication	Domestication is the process whereby a population of animals or plants, through a process of selection, becomes accustomed to human provision and control. A defining characteristic of domestication is artificial selection by humans. Humans have brought these populations under their control and care for a wide range of reasons: to produce food or valuable commodities (such as wool, cotton, or silk), for help with various types of work (such as transportation, protection, and warfare), scientific research, or simply to enjoy as companions or ornaments.
Natural selection	Natural selection is the gradual, non-random, process by which biological traits become either more or less common in a population as a function of differential reproduction of their bearers. It is a key mechanism of evolution.
	Variation exists within all populations of organisms.

Chapter 5. Evolution and Biodiversity: Origin and Diversification of Organisms

Pesticide	Pesticides are substances or mixture of substances intended for preventing, destroying, repelling or mitigating any pest. A pesticide may be a chemical, biological agent (such as a virus or bacterium), antimicrobial, disinfectant or device used against any pest. Pests include insects, plant pathogens, weeds, molluscs, birds, mammals, fish, nematodes (roundworms), and microbes that destroy property, spread disease or are vectors for disease or cause nuisance.
Allopatric speciation	Allopatric speciation is speciation that occurs when biological populations of the same species become isolated due to geographical changes such mountain building or social changes such emigration. The isolated populations then undergo genotypic and/or phenotypic divergence : (a) they become subjected to different selective pressures, (b) they independently undergo genetic drift, and (c) different mutations arise in the populations' gene pools. The separate populations over time may evolve distinctly different characteristics.
Commensalism	In ecology, commensalism is a class of relationship between two organisms where one organism benefits but the other is neutral (there is no harm or benefit). There are two other types of association: mutualism (where both organisms benefit) and parasitism (one organism benefits and the other one is harmed). Originally, the term was used to describe the use of waste food by second animals, like the carcass eaters that follow hunting animals, but wait until they have finished their meal.
Extinction	In biology and ecology, extinction is the end of an organism or of a group of organisms (taxon), normally a species. The moment of extinction is generally considered to be the death of the last individual of the species, although the capacity to breed and recover may have been lost before this point. Because a species' potential range may be very large, determining this moment is difficult, and is usually done retrospectively.
Sympatric speciation	Sympatric speciation is the proce through which new species evolve from a single ancestral species while inhabiting the same geographic region. In evolutionary biology and biogeography, sympatric and sympatry are terms referring to organisms whose ranges overlap or are even identical, so that they occur together at least in some places. If these organisms are closely related (e.g. sister species), such a distribution may be the result of sympatric speciation.

Environmental change	Environmental change is defined as a change or disturbance of the environment by natural ological processes, and is described in the following articles:

- Climate change
- Environment (biophysical)

.

Generation time	Generation time is a quantity used in population biology and demography to reflect the relative size of intervals of offspring production. Generation time usually expresses the average age of breeding females within a population. Suppose females begin breeding at age α and stop breeding (or die) at age ω, then the average age of first reproduction of a cohort of females is

$$T = \frac{\sum_{x=\alpha}^{\omega} x l(x) m(x)}{\sum_{x=\alpha}^{\omega} l(x) m(x)}$$

where $l(x)$ is the hazard function and $m(x)$ is the fecundity of females aged x.

Population size	In population genetics and population ecology, population size is the number of individual organisms in a population.

The effective population size is defined as 'the number of breeding individuals in an idealized population that would show the same amount of dispersion of allele frequencies under random genetic drift or the same amount of inbreeding as the population under consideration.' N_e is usually less than N (the absolute population size) and this has important applications in conservation genetics.

Small population size results in increased genetic drift.

Chapter 5. Evolution and Biodiversity: Origin and Diversification of Organisms

Genetically modified organism	A genetically modified organism or genetically engineered organism (GEO) is an organism whose genetic material has been altered using genetic engineering techniques. These techniques, generally known as recombinant DNA technology, use DNA molecules from different sources, which are combined into one molecule to create a new set of genes. This DNA is then transferred into an organism, giving it modified or novel genes.
Invertebrate	An invertebrate is an animal without a backbone. The group includes 97% of all animal species - all animals except those in the chordate subphylum Vertebrata (fish, amphibians, reptiles, birds, and mammals). Invertebrates form a paraphyletic group.
Species distribution	Species distribution is the manner in which a biological taxon is spatially arranged. Species distribution is not to be confused with dispersal, which is the movement of individuals away from their area of origin or from centers of high population density. A similar concept is the species range.
Global change	Global change refers to planetary-scale changes in the Earth system. The system consists of the land, oceans, atmosphere, poles, life, the planet's natural cycles and deep Earth processes. These constituent parts influence one another.
Fossil	Fossils are the preserved remains or traces of animals (also known as zoolites), plants, and other organisms from the remote past. The totality of fossils, both discovered and undiscovered, and their placement in fossiliferous (fossil-containing) rock formations and sedimentary layers (strata) is known as the fossil record. The study of fossils across geological time, how they were formed, and the evolutionary relationships between taxa (phylogeny) are some of the most important functions of the science of paleontology.

Sustainability	Sustainability is the capacity to endure. For humans, sustainability is the long-term maintenance of responsibility, which has environmental, economic, and social dimensions, and encompasses the concept of stewardship, the responsible management of resource use. In ecology, sustainability describes how biological systems remain diverse and productive over time, a necessary precondition for human well-being.
Rights	Rights are legal, social, or ethical principles of freedom or entitlement; that is, rights are the fundamental normative rules about what is allowed of people or owed to people, according to some legal system, social convention, or ethical theory. Rights are often considered fundamental to civilization, being regarded as established pillars of society and culture, and the history of social conflicts can be found in the history of each right and its development. Rights are of essential importance in such disciplines as law and ethics, especially theories of justice and deontology.
Habitat	A habitat is an ecological or environmental area that is inhabited by a particular species of animal, plant or other type of organism. It is the natural environment in which an organism lives, or the physical environment that surrounds (influences and is utilized by) a species population. Definition The term 'population' is preferred to 'organism' because, while it is possible to describe the habitat of a single black bear, it is also possible that one may not find any particular or individual bear but the grouping of bears that constitute a breeding population and occupy a certain biogeographical area.
Threatened species	Threatened species are any species (including animals, plan, fungi, etc). which are vulnerable to endangerment in the near future. The World Conservation Union (IUCN) is the foremost authority on threatened species, and trea threatened species not as a single category, but as a group of three categories, depending on the degree to which they are threatened: • Vulnerable species • Endangered species • Critically endangered species

---

Species that are threatened are sometimes characterised by the population dynamics measure of critical depensation, a mathematical measure of biomass related to population growth rate.

National park

A national park is a reserve of natural, semi-natural, or developed land that a sovereign state declares or owns. Although individual nations designate their own national parks differently , an international organization, the International Union for Conservation of Nature (IUCN), and its World Commission on Protected Areas, has defined National Parks as its category II type of protected areas.

---

1. _____ is the degree of variation of life forms within a given species, ecosystem, biome, or an entire planet. _____ is a measure of the health of ecosystems. _____ is in part a function of climate.

   a. Biological monitoring working party
   b. Biodiversity
   c. Bioreporter
   d. BioSand Filter

2. _____ are legal, social, or ethical principles of freedom or entitlement; that is, _____ are the fundamental normative rules about what is allowed of people or owed to people, according to some legal system, social convention, or ethical theory. _____ are often considered fundamental to civilization, being regarded as established pillars of society and culture, and the history of social conflicts can be found in the history of each right and its development. _____ are of essential importance in such disciplines as law and ethics, especially theories of justice and deontology.

   a. Rights ethics
   b. Ring of Gyges
   c. Synderesis
   d. Rights

3. _____ is the process whereby a population of animals or plants, through a process of selection, becomes accustomed to human provision and control. A defining characteristic of _____ is artificial selection by humans. Humans have brought these populations under their control and care for a wide range of reasons: to produce food or valuable commodities (such as wool, cotton, or silk), for help with various types of work (such as transportation, protection, and warfare), scientific research, or simply to enjoy as companions or ornaments.

   a. Domestication
   b. Bonn Convention
   c. Breeding season
   d. Buffer zone

4. In population genetics and population ecology, _____ is the number of individual organisms in a population.

   The effective _____ is defined as 'the number of breeding individuals in an idealized population that would show the same amount of dispersion of allele frequencies under random genetic drift or the same amount of inbreeding as the population under consideration.' N_e is usually less than N (the absolute _____) and this has important applications in conservation genetics.

Small _____ results in increased genetic drift.

    a. Population size
    b. Juglone
    c. Gibbons v. Ogden
    d. Global hectare

5.   _____ is the introduction of contaminants into a natural environment that causes instability, disorder, harm or discomfort to the ecosystem i.e. physical systems or living organisms. _____ can take the form of chemical substances or energy, such as noise, heat or light. Pollutants, the components of _____, can be either foreign substances/energies or naturally occurring contaminants.

    a. Basic precipitation
    b. Bioaccumulation
    c. Bioconcentration factor
    d. Pollution

---

1. b
2. d
3. a
4. a
5. d

You can take the complete Chapter Practice Test

for Chapter 5. Evolution and Biodiversity: Origin and Diversification of Organisms
on all key terms, persons, places, and concepts.

Online 99 Cents

http://www.epub89.16.20190.5.cram101.com/

Use www.Cram101.com for all your study needs

including Cram101's online interactive problem solving labs in chemistry, statistics, mathematics, and more.

Ecological succession

Secondary succession

Invertebrate

Air pollution

Pollution

Biosphere

Complexity

Biome

Population ecology

Biodiversity

Biodiversity hotspot

Ecology

Hotspot

Green Revolution

Population density

Population size

Density

Carrying capacity

Sex ratio

Ratio

Scientific method

Predation

Keystone species

Parasitoid

Porcupine

Alkali

Commensalism

Coral reef

Primary

Primary succession

Habitat

Island biogeography

Species richness

Biogeography

Sustainability

Ecological succession	Ecological succession, is the phenomenon or process by which an ecological community undergoes more or less orderly and predictable changes following disturbance or initial colonization of new habitat. Succession was among the first theories advanced in ecology and the study of succession remains at the core of ecological science. Succession may be initiated either by formation of new, unoccupied habitat (e.g., a lava flow or a severe landslide) or by some form of disturbance (e.g. fire, severe windthrow, logging) of an existing community.
Secondary succession	Secondary succession is one of the two types of ecological succession of plant life. As opposed to the first, primary succession, secondary succession is a process started by an event (e.g. forest fire, harvesting, hurricane) that reduces an already established ecosystem (e.g. a forest or a wheat field) to a smaller population of species, and as such secondary succession occurs on preexisting soil whereas primary succession usually occurs in a place lacking soil.
Invertebrate	An invertebrate is an animal without a backbone. The group includes 97% of all animal species - all animals except those in the chordate subphylum Vertebrata (fish, amphibians, reptiles, birds, and mammals). Invertebrates form a paraphyletic group.
Air pollution	Air pollution is the introduction of chemicals, particulate matter, or biological materials that cause harm or discomfort to humans or other living organisms, or cause damage to the natural environment or built environment, into the atmosphere. The atmosphere is a complex dynamic natural gaseous system that is essential to support life on planet Earth. Stratospheric ozone depletion due to air pollution has long been recognized as a threat to human health as well as to the Earth's ecosystems.
Pollution	Pollution is the introduction of contaminants into a natural environment that causes instability, disorder, harm or discomfort to the ecosystem i.e. physical systems or living organisms. Pollution can take the form of chemical substances or energy, such as noise, heat or light. Pollutants, the components of pollution, can be either foreign substances/energies or naturally occurring contaminants.

Chapter 6. Population and Community Ecology: Distribution and Abundance of Species

Biosphere	The biosphere is the global sum of all ecosystems. It can also be called the zone of life on Earth, a closed (apart from solar and cosmic radiation) and self-regulating system. From the broadest biophysiological point of view, the biosphere is the global ecological system integrating all living beings and their relationships, including their interaction with the elements of the lithosphere, hydrosphere and atmosphere.
Complexity	In general usage, complexity tends to be used to characterize something with many parts in intricate arrangement. The study of these complex linkages is the main goal of complex systems theory. In science there are at this time a number of approaches to characterizing complexity, many of which are reflected in this article.
Biome	Biomes are climatically and geographically defined as similar climatic conditions on the Earth, such as communities of plants, animals, and soil organisms, and are often referred to as ecosystems. Some parts of the earth have more or less the same kind of abiotic and biotic factors spread over a large area, creating a typical ecosystem over that area. Such major ecosystems are termed as biomes.
Population ecology	Population ecology is a major sub-field of ecology that deals with the dynamics of species populations and how these populations interact with the environment. It is the study of how the population sizes of species living together in groups change over time and space. The development of population ecology owes much to demography and actuarial life tables.
Biodiversity	Biodiversity is the degree of variation of life forms within a given species, ecosystem, biome, or an entire planet. Biodiversity is a measure of the health of ecosystems. Biodiversity is in part a function of climate.
Biodiversity hotspot	A biodiversity hotspot is a biogeographic region with a significant reservoir of biodiversity that is under threat from humans. The concept of biodiversity hotspots was originated by Norman Myers in two articles in 'The Environmentalist' (1988 ' 1990), revised after thorough analysis by Myers and others in 'Hotspots: Earth's Biologically Richest and Most Endangered Terrestrial Ecoregions'.

To qualify as a biodiversity hotspot on Myers 2000 edition of the hotspot-map, a region must meet two strict criteria: it must contain at least 0.5% or 1,500 species of vascular plants as endemics, and it has to have lost at least 70% of its primary vegetation.

Ecology	Ecology is the scientific study of the relations that living organisms have with respect to each other and their natural environment. Variables of interest to ecologists include the composition, distribution, amount (biomass), number, and changing states of organisms within and among ecosystems. Ecosystems are hierarchical systems that are organized into a graded series of regularly interacting and semi-independent parts (e.g., species) that aggregate into higher orders of complex integrated wholes (e.g., communities).
Hotspot	The places known as hotspots or hot spots in geology are volcanic regions thought to be fed by underlying mantle that is anomalously hot compared with the mantle elsewhere. They may be on, near to, or far from tectonic plate boundaries. There are two hypotheses to explain them.
Green Revolution	Green Revolution refers to a series of research, development, and technology transfer initiatives, occurring between the 1940s and the late 1970s, that increased agriculture production around the world, beginning most markedly in the late 1960s.
	The initiatives, led by Norman Borlaug, the 'Father of the Green Revolution' credited with saving over a billion people from starvation, involved the development of high-yielding varieties of cereal grains, expansion of irrigation infrastructure, modernization of management techniques, distribution of hybridized seeds, synthetic fertilizers, and pesticides to farmers.
	The term 'Green Revolution' was first used in 1968 by former United States Agency for International Development (USAID) director William Gaud, who noted the spread of the new technologies and said,
	These and other developments in the field of agriculture contain the makings of a new revolution.
Population density	Population density is a measurement of population per unit area or unit volume. It is frequently applied to living organisms, and particularly to humans. It is a key geographic term.

Population size	In population genetics and population ecology, population size is the number of individual organisms in a population.
	The effective population size is defined as 'the number of breeding individuals in an idealized population that would show the same amount of dispersion of allele frequencies under random genetic drift or the same amount of inbreeding as the population under consideration.' N_e is usually less than N (the absolute population size) and this has important applications in conservation genetics.
	Small population size results in increased genetic drift.
Density	The mass density is defined as its mass per unit volume. The symbol most often used for density is ρ. In some cases (for instance, in the United States oil and gas industry), density is also defined as its weight per unit volume; although, this quantity is more properly called specific weight.
Carrying capacity	The carrying capacity of a biological species in an environment is the maximum population size of the species that the environment can sustain indefinitely, given the food, habitat, water and other necessities available in the environment. In population biology, carrying capacity is defined as the environment's maximal load, which is different from the concept of population equilibrium.
	For the human population, more complex variables such as sanitation and medical care are sometimes considered as part of the necessary establishment.
Sex ratio	Sex ratio is the ratio of males to females in a population. The primary sex ratio is the ratio at the time of conception, secondary sex ratio is the ratio at time of birth, and tertiary sex ratio is the ratio of mature organisms.

The human sex ratio is of particular interest to anthropologists and demographers. In humans the secondary sex ratio is commonly assumed to be 105 boys to 100 girls (which sometimes is shortened to 'a ratio of 105'). In human societies, however, sex ratios at birth may be considerably skewed by natural reasons such as the age of mother at birth, and unnatural reasons such as sex-selective abortion. The CIA estimates that the current world wide sex ratio at birth is 107 boys to 100 girls. The value for the entire world population is 101 males to 100 females.

Ratio

In mathematics, a ratio is a relationship between two numbers of the same kind (e.g., objects, persons, students, spoonfuls, units of whatever identical dimension), usually expressed as 'a to b' or a:b, sometimes expressed arithmetically as a dimensionless quotient of the two which explicitly indicates how many times the first number contains the second (not necessarily an integer). In layman's terms a ratio represents, simply, for every amount of one thing, how much there is of another thing. For example, suppose I have 10 pairs of socks for every pair of shoes then the ratio of shoes:socks would be 1:10 and the ratio of socks:shoes would be 10:1

Notation and terminology
The ratio of numbers A and B can be expressed as:

- the ratio of A to B
- A is to B
- A:B
- A rational number which is the quotient of A divided by B

The numbers A and B are sometimes called terms with A being the antecedent and B being the consequent.

Scientific method

Scientific method refers to a body of techniques for investigating phenomena, acquiring new knowledge, or correcting and integrating previous knowledge. To be termed scientific, a method of inquiry must be based on gathering empirical and measurable evidence subject to specific principles of reasoning. The Oxford English Dictionary says that scientific method is: 'a method or procedure that has characterized natural science since the 17th century, consisting in systematic observation, measurement, and experiment, and the formulation, testing, and modification of hypotheses.'

	The chief characteristic which distinguishes a scientific method of inquiry from other methods of acquiring knowledge is that scientists seek to let reality speak for itself, and contradict their theories about it when those theories are incorrect, i. e., falsifiability.
Predation	In ecology, predation describes a biological interaction where a predator (an organism that is hunting) feeds on its prey (the organism that is attacked). Predators may or may not kill their prey prior to feeding on them, but the act of predation always results in the death of its prey and the eventual absorption of the prey's tissue through consumption. Other categories of consumption are herbivory (eating parts of plants) and detritivory, the consumption of dead organic material (detritus).
Keystone species	A keystone species is a species that has a disproportionately large effect on its environment relative to its abundance. Such species play a critical role in maintaining the structure of an ecological community, affecting many other organisms in an ecosystem and helping to determine the types and numbers of various other species in the community. The role that a keystone species plays in its ecosystem is analogous to the role of a keystone in an arch.
Parasitoid	A parasitoid is an organism that spends a significant portion of its life history attached to or within a single host organism in a relationship that is in essence parasitic; unlike a true parasite, however, it ultimately sterilises or kills, and sometimes consumes, the host. Thus parasitoids are similar to typical parasites except in the more dire prognosis for the host. Definitions and distinctions The term parasitoid was coined in 1913 by the German writer O. M. Reuter to describe the strategy in which, during its development, the parasite lives in or on the body of a single host individual, eventually killing that host, the adult parasitoid being free-living.
Porcupine	Porcupines are rodents with a coat of sharp spines, or quills, that defend or camouflage them from predators. They are indigenous to the Americas, southern Asia, and Africa. Porcupines are the third largest of the rodents, behind the capybara and the beaver. Most porcupines are about 25-36 in (63-91 cm) long, with an 8-10 in (20-25 cm) long tail.

Alkali	In chemistry, an alkali is a basic, ionic salt of an alkali metal or alkaline earth metal element. Some authors also define an alkali as a base that dissolves in water. A solution of a soluble base has a pH greater than 7. The adjective alkaline is commonly used in English as a synonym for base, especially for soluble bases.
Commensalism	In ecology, commensalism is a class of relationship between two organisms where one organism benefits but the other is neutral (there is no harm or benefit). There are two other types of association: mutualism (where both organisms benefit) and parasitism (one organism benefits and the other one is harmed). Originally, the term was used to describe the use of waste food by second animals, like the carcass eaters that follow hunting animals, but wait until they have finished their meal.
Coral reef	Coral reefs are underwater structures made from calcium carbonate secreted by corals. Corals are colonies of tiny living animals found in marine waters that contain few nutrients. Most coral reefs are built from stony corals, which in turn consist of polyps that cluster in groups.
Primary	A primary (or gravitational primary) is the main physical body of a gravitationally bound, multi-object system. This body contributes most of the mass of that system and will generally be located near its center of mass. In the solar system, the Sun is the primary for all objects that orbit around it.
Primary succession	Primary succession is one of two types of biological and ecological succession of plant life, occurring in an environment in which new substrate devoid of vegetation and usually lacking soil, such as a lava flow or area left from retreated glacier, is deposited. In other words, it is the gradual growth of an ecosystem over a longer period of time. In contrast, secondary succession occurs on substrate that previously supported vegetation before an ecological disturbance such as forest fire, tsunami, flood, destroyed the plant life.

Chapter 6. Population and Community Ecology: Distribution and Abundance of Species

Habitat	A habitat is an ecological or environmental area that is inhabited by a particular species of animal, plant or other type of organism. It is the natural environment in which an organism lives, or the physical environment that surrounds (influences and is utilized by) a species population. Definition The term 'population' is preferred to 'organism' because, while it is possible to describe the habitat of a single black bear, it is also possible that one may not find any particular or individual bear but the grouping of bears that constitute a breeding population and occupy a certain biogeographical area.
Island biogeography	Island biogeography is a field within biogeography that attempts to establish and explain the factors that affect the species richness of natural communities. The theory was developed to explain species richness of actual islands. It has since been extended to mountains surrounded by deserts, lakes surrounded by dry land, fragmented forest and even natural habitats surrounded by human-altered landscapes.
Species richness	Species richness is the number of different species in a given area. It is represented in equation form as S. Species richness is the fundamental unit in which to assess the homogeneity of an environment. Typically, species richness is used in conservation studies to determine the sensitivity of ecosystems and their resident species. The actual number of species calculated alone is largely an arbitrary number. These studies, therefore, often develop a rubric or measure for valuing the species richness number(s) or adopt one from previous studies on similar ecosystems.
Biogeography	Biogeography is the study of the distribution of species (biology), organisms, and ecosystems in space and through geological time. Organisms and biological communities vary in a highly regular fashion along geographic gradients of latitude, elevation, isolation and habitat area.

Knowledge of spatial variation in the numbers and types of organisms is as vital to us today as it was to our early human ancestors, as we adapt to heterogeneous but geographically predictable environments.

| Sustainability | Sustainability is the capacity to endure. For humans, sustainability is the long-term maintenance of responsibility, which has environmental, economic, and social dimensions, and encompasses the concept of stewardship, the responsible management of resource use. In ecology, sustainability describes how biological systems remain diverse and productive over time, a necessary precondition for human well-being. |

---

1. A _____ is an ecological or environmental area that is inhabited by a particular species of animal, plant or other type of organism. It is the natural environment in which an organism lives, or the physical environment that surrounds (influences and is utilized by) a species population.

    Definition
    The term 'population' is preferred to 'organism' because, while it is possible to describe the _____ of a single black bear, it is also possible that one may not find any particular or individual bear but the grouping of bears that constitute a breeding population and occupy a certain biogeographical area.

    a. Habitat
    b. Habitat destruction
    c. Landscape limnology
    d. Levels of organization

2. In population genetics and population ecology, _____ is the number of individual organisms in a population.

    The effective _____ is defined as 'the number of breeding individuals in an idealized population that would show the same amount of dispersion of allele frequencies under random genetic drift or the same amount of inbreeding as the population under consideration.' N_e is usually less than N (the absolute _____) and this has important applications in conservation genetics.

    Small _____ results in increased genetic drift.

    a. genetic diversity
    b. Juglone
    c. Population size
    d. Rank-size distribution

3. The _____ is the global sum of all ecosystems. It can also be called the zone of life on Earth, a closed (apart from solar and cosmic radiation) and self-regulating system. From the broadest biophysiological point of view, the _____ is the global ecological system integrating all living beings and their relationships, including their interaction with the elements of the lithosphere, hydrosphere and atmosphere.

---

a. Bjerrum plot
b. Bottom crawler
c. Biosphere
d. Brine pool

4. _____, is the phenomenon or process by which an ecological community undergoes more or less orderly and predictable changes following disturbance or initial colonization of new habitat. Succession was among the first theories advanced in ecology and the study of succession remains at the core of ecological science. Succession may be initiated either by formation of new, unoccupied habitat (e.g., a lava flow or a severe landslide) or by some form of disturbance (e.g. fire, severe windthrow, logging) of an existing community.

a. Odyssey
b. Ecological succession
c. Absent-minded professor
d. Abstract object

5. _____ is one of the two types of ecological succession of plant life. As opposed to the first, primary succession, _____ is a process started by an event (e.g. forest fire, harvesting, hurricane) that reduces an already established ecosystem (e.g. a forest or a wheat field) to a smaller population of species, and as such _____ occurs on preexisting soil whereas primary succession usually occurs in a place lacking soil.

a. Natural environment
b. Juglone
c. Gibbons v. Ogden
d. Secondary succession

---

1. a
2. c
3. c
4. b
5. d

You can take the complete Chapter Practice Test

for Chapter 6. Population and Community Ecology: Distribution and Abundance of Species
on all key terms, persons, places, and concepts.

Online 99 Cents

http://www.epub89.16.20190.6.cram101.com/

Use www.Cram101.com for all your study needs

including Cram101's online interactive problem solving labs in chemistry, statistics, mathematics, and more.

Human Populations: Patterns and Processes of Human Population Growth

Carrying capacity

Green Revolution

Demography

Thomas Garnett

Total fertility rate

Infant mortality

Population pyramid

Environmental justice

Population momentum

Momentum

Demographic transition

Natural resource

Sustainable development

Air pollution

Pollution

Population size

Ecological footprint

Efficiency

Equation

| | Millennium Ecosystem Assessment |

| | Sustainability |

| | Biodiversity |

| | Biodiversity hotspot |

| | Hotspot |

| | MONK |

Carrying capacity	The carrying capacity of a biological species in an environment is the maximum population size of the species that the environment can sustain indefinitely, given the food, habitat, water and other necessities available in the environment. In population biology, carrying capacity is defined as the environment's maximal load, which is different from the concept of population equilibrium.
	For the human population, more complex variables such as sanitation and medical care are sometimes considered as part of the necessary establishment.
Green Revolution	Green Revolution refers to a series of research, development, and technology transfer initiatives, occurring between the 1940s and the late 1970s, that increased agriculture production around the world, beginning most markedly in the late 1960s.
	The initiatives, led by Norman Borlaug, the 'Father of the Green Revolution' credited with saving over a billion people from starvation, involved the development of high-yielding varieties of cereal grains, expansion of irrigation infrastructure, modernization of management techniques, distribution of hybridized seeds, synthetic fertilizers, and pesticides to farmers.
	The term 'Green Revolution' was first used in 1968 by former United States Agency for International Development (USAID) director William Gaud, who noted the spread of the new technologies and said,
	These and other developments in the field of agriculture contain the makings of a new revolution.
Demography	Demography is the statistical study of human populations and sub-populations. It can be a very general science that can be applied to any kind of dynamic human population, that is, one that changes over time or space . It encompasses the study of the size, structure, and distribution of these populations, and spatial and/or temporal changes in them in response to birth, migration, aging and death.

| Thomas Garnett | Thomas Garnett (1766-1802) was an English physician and natural philosopher.

Life
He was born 21 April 1766 at Casterton in Westmoreland, where his father had a small landed property. After attending a local school he was at the age of fifteen articled at his own request to the celebrated John Dawson (surgeon) of Sedbergh, Yorkshire, surgeon and mathematician. |
| --- | --- |
| Total fertility rate | The total fertility rate of a population is the average number of children that would be born to a woman over her lifetime if (1) she were to experience the exact current age-specific fertility rates (ASFRs) through her lifetime, and (2) she were to survive from birth through the end of her reproductive life. It is obtained by summing the single-year age-specific rates at a given time. |
| Infant mortality | Infant mortality is defined as the number of infant deaths (one year of age or younger) per 1000 live births. Traditionally, the most common cause worldwide was dehydration from diarrhea. However, the spreading information about Oral Re-hydration Solution (a mixture of salts, sugar, and water) to mothers around the world has decreased the rate of children dying from dehydration. |
| Population pyramid | A population pyramid, is a graphical illustration that shows the distribution of various age groups in a human population (typically that of a country or region of the world), which ideally forms the shape of a pyramid when the region is healthy. It is also used in Ecology to determine the overall age distribution of a population; an indication of the reproductive capabilities and likelihood of the continuation of a species.

It typically consists of two back-to-back bar graphs, with the population plotted on the X-axis and age on the Y-axis, one showing the number of males and one showing females in a particular population in five-year age groups (also called cohorts). |
| Environmental justice | Environmental justice is 'the fair treatment and meaningful involvement of all people regardless of race, color, sex, national origin, or income with respect to the development, implementation and enforcement of environmental laws, regulations, and policies.' In the words of Bunyan Bryant, 'Environmental justice is served when people can realize their highest potential.' |

Environmental justice emerged as a concept in the United States in the early 1980s; its proponents generally view the environment as encompassing 'where we live, work, and play' (sometimes 'pray' and 'learn' are also included) and seek to redress inequitable distributions of environmental burdens (pollution, industrial facilities, crime, etc).. Root causes of environmental injustices include 'institutionalized racism; the co-modification of land, water, energy and air; unresponsive, unaccountable government policies and regulation; and lack of resources and power in affected communities.'

Definition
The United States Environmental Protection Agency defines as follows:

'Environmental Justice is the fair treatment and meaningful involvement of all people regardless of race, color, national origin, or income with respect to the development, implementation, and enforcement of environmental laws, regulations, and policies. EPA has this goal for all communities and persons across this Nation.

Population momentum	Population momentum refers to population growth at the national level which would occur even if levels of childbearing immediately declined to replacement level. For countries with above-replacement fertility (greater than 2.1 children per woman), population momentum represents natural increase to the population. For below-replacement countries, momentum corresponds to a population decline.
Momentum	In classical mechanics, linear momentum or translational momentum is the product of the mass and velocity of an object: $$\mathbf{p} \equiv m\mathbf{v}.$$ Like velocity, linear momentum is a vector quantity, possessing a direction as well as a magnitude. Linear momentum is also a conserved quantity, meaning that if a closed system is not affected by external forces, its total linear momentum cannot change.

Demographic transition	The demographic transition is the transition from high birth and death rates to low birth and death rates as a country develops from a pre-industrial to an industrialized economic system. The theory is based on an interpretation of demographic history developed in 1929 by the American demographer Warren Thompson (1887-1973). Thompson observed changes, or transitions, in birth and death rates in industrialized societies over the previous 200 years.
Natural resource	Natural resources occur naturally within environments that exist relatively undisturbed by mankind, in a natural form. A natural resource is often characterized by amounts of biodiversity and geodiversity existent in various ecosystems.
	Natural resources are derived from the environment.
Sustainable development	Sustainable development is a pattern of growth in which resource use aims to meet human needs while preserving the environment so that these needs can be met not only in the present, but also for generations to come (sometimes taught as ELF-Environment, Local people, Future). The term sustainable development was used by the Brundtland Commission which coined what has become the most often-quoted definition of sustainable development as development that 'meets the needs of the present without compromising the ability of future generations to meet their own needs.'
	Sustainable development ties together concern for the carrying capacity of natural systems with the social challenges facing humanity. As early as the 1970s 'sustainability' was employed to describe an economy 'in equilibrium with basic ecological support systems.' Ecologists have pointed to The Limits to Growth, and presented the alternative of a 'steady state economy' in order to address environmental concerns.
Air pollution	Air pollution is the introduction of chemicals, particulate matter, or biological materials that cause harm or discomfort to humans or other living organisms, or cause damage to the natural environment or built environment, into the atmosphere.

	The atmosphere is a complex dynamic natural gaseous system that is essential to support life on planet Earth. Stratospheric ozone depletion due to air pollution has long been recognized as a threat to human health as well as to the Earth's ecosystems.
Pollution	Pollution is the introduction of contaminants into a natural environment that causes instability, disorder, harm or discomfort to the ecosystem i.e. physical systems or living organisms. Pollution can take the form of chemical substances or energy, such as noise, heat or light. Pollutants, the components of pollution, can be either foreign substances/energies or naturally occurring contaminants.
Population size	In population genetics and population ecology, population size is the number of individual organisms in a population.
	The effective population size is defined as 'the number of breeding individuals in an idealized population that would show the same amount of dispersion of allele frequencies under random genetic drift or the same amount of inbreeding as the population under consideration.' N_e is usually less than N (the absolute population size) and this has important applications in conservation genetics.
	Small population size results in increased genetic drift.
Ecological footprint	The ecological footprint is a measure of human demand on the Earth's ecosystems. It is a standardized measure of demand for natural capital that may be contrasted with the planet's ecological capacity to regenerate. It represents the amount of biologically productive land and sea area necessary to supply the resources a human population consumes, and to assimilate associated waste.
Efficiency	Efficiency in general describes the extent to which time or effort is well used for the intended task or purpose. It is often used with the specific purpose of relaying the capability of a specific application of effort to produce a specific outcome effectively with a minimum amount or quantity of waste, expense, or unnecessary effort. 'Efficiency' has widely varying meanings in different disciplines.

Equation	An equation is a mathematical statement that asserts the equality of two expressions. In modern notation, this is written by placing the expressions on either side of an equals sign (=), for example
	$$x + 3 = 5$$
	asserts that x+3 is equal to 5. The = symbol was invented by Robert Recorde (1510-1558), who considered that nothing could be more equal than parallel straight lines with the same length.
	Knowns and unknowns
	Equations often express relationships between given quantities, the knowns, and quantities yet to be determined, the unknowns.
Millennium Ecosystem Assessment	The Millennium Ecosystem Assessment, released in 2005, is an international synthesis by over 1000 of the world's leading biological scientists that analyses the state of the Earth's ecosystems and provides summaries and guidelines for decision-makers. It concludes that human activity is having a significant and escalating impact on the biodiversity of world ecosystems, reducing both their resilience and biocapacity. The report refers to natural systems as humanity's 'life-support system', providing essential 'ecosystem services'.
Sustainability	Sustainability is the capacity to endure. For humans, sustainability is the long-term maintenance of responsibility, which has environmental, economic, and social dimensions, and encompasses the concept of stewardship, the responsible management of resource use. In ecology, sustainability describes how biological systems remain diverse and productive over time, a necessary precondition for human well-being.
Biodiversity	Biodiversity is the degree of variation of life forms within a given species, ecosystem, biome, or an entire planet. Biodiversity is a measure of the health of ecosystems. Biodiversity is in part a function of climate.
Biodiversity hotspot	A biodiversity hotspot is a biogeographic region with a significant reservoir of biodiversity that is under threat from humans.

The concept of biodiversity hotspots was originated by Norman Myers in two articles in 'The Environmentalist' (1988 ' 1990), revised after thorough analysis by Myers and others in 'Hotspots: Earth's Biologically Richest and Most Endangered Terrestrial Ecoregions'.

To qualify as a biodiversity hotspot on Myers 2000 edition of the hotspot-map, a region must meet two strict criteria: it must contain at least 0.5% or 1,500 species of vascular plants as endemics, and it has to have lost at least 70% of its primary vegetation.

Hotspot

The places known as hotspots or hot spots in geology are volcanic regions thought to be fed by underlying mantle that is anomalously hot compared with the mantle elsewhere. They may be on, near to, or far from tectonic plate boundaries. There are two hypotheses to explain them.

MONK

MONK is a Monte Carlo software package for simulating nuclear processes, particularly for the purpose of determining the neutron multiplication factor, or k-effective, of a system. It is owned by Serco Assurance. It is used primarily to judge whether or not nuclear systems are critical.

---

1. _____ refers to a series of research, development, and technology transfer initiatives, occurring between the 1940s and the late 1970s, that increased agriculture production around the world, beginning most markedly in the late 1960s.

   The initiatives, led by Norman Borlaug, the 'Father of the _____' credited with saving over a billion people from starvation, involved the development of high-yielding varieties of cereal grains, expansion of irrigation infrastructure, modernization of management techniques, distribution of hybridized seeds, synthetic fertilizers, and pesticides to farmers.

   The term '_____' was first used in 1968 by former United States Agency for International Development (USAID) director William Gaud, who noted the spread of the new technologies and said,

   These and other developments in the field of agriculture contain the makings of a new revolution.

   a. Juglone
   b. Green Revolution
   c. Krakatoa
   d. Refuse Act

2. _____ refers to population growth at the national level which would occur even if levels of childbearing immediately declined to replacement level. For countries with above-replacement fertility (greater than 2.1 children per woman), _____ represents natural increase to the population. For below-replacement countries, momentum corresponds to a population decline.

   a. Nomad
   b. Juglone
   c. Population momentum
   d. Environmental psychology

3. The _____ of a biological species in an environment is the maximum population size of the species that the environment can sustain indefinitely, given the food, habitat, water and other necessities available in the environment. In population biology, _____ is defined as the environment's maximal load, which is different from the concept of population equilibrium.

For the human population, more complex variables such as sanitation and medical care are sometimes considered as part of the necessary establishment.

a. Carrying capacity
b. Gibbons v. Ogden
c. Krakatoa
d. Refuse Act

4. _____ is the statistical study of human populations and sub-populations. It can be a very general science that can be applied to any kind of dynamic human population, that is, one that changes over time or space . It encompasses the study of the size, structure, and distribution of these populations, and spatial and/or temporal changes in them in response to birth, migration, aging and death.

a. Demography
b. Human ecology
c. Human geography
d. Natural resource management

5. An _____ is a mathematical statement that asserts the equality of two expressions. In modern notation, this is written by placing the expressions on either side of an equals sign (=), for example

$$x + 3 = 5$$

asserts that x+3 is equal to 5. The = symbol was invented by Robert Recorde (1510-1558), who considered that nothing could be more equal than parallel straight lines with the same length.

Knowns and unknowns
_____s often express relationships between given quantities, the knowns, and quantities yet to be determined, the unknowns.

a. Identity
b. Inequality
c. Equation
d. Unary operation

---

1. b
2. c
3. a
4. a
5. c

You can take the complete Chapter Practice Test

for Chapter 7. Human Populations: Patterns and Processes of Human Population Growth
on all key terms, persons, places, and concepts.

Online 99 Cents

http://www.epub89.16.20190.7.cram101.com/

Use www.Cram101.com for all your study needs

including Cram101's online interactive problem solving labs in chemistry, statistics, mathematics, and more.

Solar System

Carrying capacity

Asthenosphere

Convection

Lithosphere

Magma

Tectonics

Pangaea

Plate tectonics

Seafloor spreading

Subduction

Biodiversity

Biodiversity hotspot

Geologic time scale

Hotspot

Earthquake

Himalayas

Transform fault

Habitat

Chapter 8. Earth`s Resources: Geologic Processes, Soil, and Minerals

_____ | Mineral _____

_____ | Rock cycle _____

_____ | Global warming _____

_____ | Anthracite _____

_____ | Basalt _____

_____ | Extrusive _____

_____ | Granite _____

_____ | Igneous rock _____

_____ | Obsidian _____

_____ | Sediment _____

_____ | Sedimentary rock _____

_____ | Pollution _____

_____ | Sandstone _____

_____ | Weathering _____

_____ | Feldspar _____

_____ | Air pollution _____

_____ | Primary _____

_____ | Ecosystem services _____

_____ | Erosion _____

_____ | Parent material

_____ | Topography

_____ | Topsoil

_____ | Soil texture

_____ | Green Revolution

_____ | Porosity

_____ | Cation exchange capacity

_____ | Landfill

_____ | Bauxite

_____ | Natural resource

_____ | Mount Pinatubo

_____ | Placer mining

_____ | Surface mining

_____ | Taiga

_____ | Sustainability

Solar System	The Solar System consists of the Sun and the astronomical objects gravitationally bound in orbit around it, all of which formed from the collapse of a giant molecular cloud approximately 4.6 billion years ago. The vast majority of the system's ma (well over 99%) is in the Sun. Of the many objects that orbit the Sun, most of the ma is contained within eight relatively solitary planets[e] whose orbits are almost circular and lie within a nearly flat disc called the ecliptic plane.
Carrying capacity	The carrying capacity of a biological species in an environment is the maximum population size of the species that the environment can sustain indefinitely, given the food, habitat, water and other necessities available in the environment. In population biology, carrying capacity is defined as the environment's maximal load, which is different from the concept of population equilibrium.
	For the human population, more complex variables such as sanitation and medical care are sometimes considered as part of the necessary establishment.
Asthenosphere	The asthenosphere is the highly viscous mechanically weak ductilely-deforming region of the upper mantle of the Earth. It lies below the lithosphere, at depths between 100 and 200 km (~ 62 and 124 miles) below the surface, but perhaps extending as deep as 700 km (~ 435 miles).
	Characteristics
	The asthenosphere is a portion of the upper mantle just below the lithosphere that is involved in plate tectonic movements and isostatic adjustments.
Convection	Convection is the concerted, collective movement of ensembles of molecules within fluids (i.e. liquids, gases) and rheids. Convection of mass cannot take place in solids, since neither bulk current flows nor significant diffusion can take place in solids. Diffusion of heat can take place in solids, but is referred to separately in that case as heat conduction.
Lithosphere	The lithosphere is the rigid outermost shell of a rocky planet. On Earth, it comprises the crust and the portion of the upper mantle that behaves elastically on time scales of thousands of years or greater.
	Earth's lithosphere

In the Earth the lithosphere includes the crust and the uppermost mantle, which constitute the hard and rigid outer layer of the Earth.

Magma

Magma is a mixture of molten or semi molten rock, volatiles and solids that is found beneath the surface of the Earth, and is expected to exist on other terrestrial planets. Besides molten rock, magma may also contain suspended crystals and dissolved gas and sometimes also gas bubbles. Magma often collects in magma chambers that may feed a volcano or turn into a pluton.

Tectonics

Tectonics is a field of study within geology concerned generally with the structures within the lithosphere of the Earth (or other planets) and particularly with the forces and movements that have operated in a region to create these structures.

Tectonics is concerned with the orogenies and tectonic development of cratons and tectonic terranes as well as the earthquake and volcanic belts which directly affect much of the global population. Tectonic studies are also important for understanding erosion patterns in geomorphology and as guides for the economic geologist searching for petroleum and metallic ores.

Pangaea

Pangaea was the supercontinent that existed during the Paleozoic and Mesozoic eras about 250 million years ago, before the component continents were separated into their current configuration.

The name was coined during a 1926 symposium discussing Alfred Wegener's theory of continental drift. In his book The Origin of Continents and Oceans (Die Entstehung der Kontinente und Ozeane) first published in 1915, he postulated that all the continents had at one time formed a single supercontinent which he called the 'Urkontinent', before later breaking up and drifting to their present locations.

Chapter 8. Earth`s Resources: Geologic Processes, Soil, and Minerals

Plate tectonics	Plate tectonics is a scientific theory which describes the large scale motions of Earth's lithosphere. The theory builds on the older concepts of continental drift, developed during the first decades of the 20th century (one of the most famous advocates was Alfred Wegener), and was accepted by the majority of the geoscientific community when the concepts of seafloor spreading were developed in the late 1950s and early 1960s. The lithosphere is broken up into what are called tectonic plates.
Seafloor spreading	Seafloor spreading is a process that occurs at mid-ocean ridges, where new oceanic crust is formed through volcanic activity and then gradually moves away from the ridge. Seafloor spreading helps explain continental drift in the theory of plate tectonics. Earlier theories (e.g., by Alfred Wegener and Alexander du Toit) of continental drift were that continents 'plowed' through the sea.
Subduction	In geology, subduction is the process that takes place at convergent boundaries by which one tectonic plate moves under another tectonic plate, sinking into the Earth's mantle, as the plates converge. These 3D regions of mantle downwellings are known as 'Subduction Zones'. A subduction zone is an area on Earth where two tectonic plates move towards one another and one slides under the other.
Biodiversity	Biodiversity is the degree of variation of life forms within a given species, ecosystem, biome, or an entire planet. Biodiversity is a measure of the health of ecosystems. Biodiversity is in part a function of climate.
Biodiversity hotspot	A biodiversity hotspot is a biogeographic region with a significant reservoir of biodiversity that is under threat from humans. The concept of biodiversity hotspots was originated by Norman Myers in two articles in 'The Environmentalist' (1988 ' 1990), revised after thorough analysis by Myers and others in 'Hotspots: Earth's Biologically Richest and Most Endangered Terrestrial Ecoregions'.

To qualify as a biodiversity hotspot on Myers 2000 edition of the hotspot-map, a region must meet two strict criteria: it must contain at least 0.5% or 1,500 species of vascular plants as endemics, and it has to have lost at least 70% of its primary vegetation.

Geologic time scale

The geologic time scale provides a system of chronologic measurement relating stratigraphy to time that is used by geologists, paleontologists and other earth scientists to describe the timing and relationships between events that have occurred during the history of the Earth. The table of geologic time spans presented here agrees with the dates and nomenclature proposed by the International Commission on Stratigraphy, and uses the standard color codes of the United States Geological Survey.

Evidence from radiometric dating indicates that the Earth is about 4.570 billion years old.

Hotspot

The places known as hotspots or hot spots in geology are volcanic regions thought to be fed by underlying mantle that is anomalously hot compared with the mantle elsewhere. They may be on, near to, or far from tectonic plate boundaries. There are two hypotheses to explain them.

Earthquake

An earthquake is the result of a sudden release of energy in the Earth's crust that creates seismic waves. The seismicity or seismic activity of an area refers to the frequency, type and size of earthquakes experienced over a period of time. Earthquakes are measured using observations from seismometers.

Himalayas

The Himalayas, is a mountain range immediately at the north of the Indian subcontinent. By extension, it is also the name of a massive mountain system that includes the Karakoram, the Hindu Kush, and other, lesser, ranges that extend out from the Pamir Knot.

Together, the Himalayan mountain system is the world's highest, and home to the world's highest peaks, the Eight-thousanders, which include Mount Everest and K2. To comprehend the enormous scale of this mountain range, consider that Aconcagua, in the Andes, at 6,962 metres (22,841 ft) is the highest peak outside Asia, whereas the Himalayan system includes over 100 mountains exceeding 7,200 m (23,600 ft).

Transform fault	A transform fault, also known as conservative plate boundary since these faults neither create nor destroy lithosphere. This is a type of fault whose relative motion is predominantly horizontal in either sinistral or dextral direction. Furthermore, transform faults end abruptly and are connected on both ends to other faults, ridges, or subduction zones.
Habitat	A habitat is an ecological or environmental area that is inhabited by a particular species of animal, plant or other type of organism. It is the natural environment in which an organism lives, or the physical environment that surrounds (influences and is utilized by) a species population. Definition The term 'population' is preferred to 'organism' because, while it is possible to describe the habitat of a single black bear, it is also possible that one may not find any particular or individual bear but the grouping of bears that constitute a breeding population and occupy a certain biogeographical area.
Mineral	A mineral is a naturally occurring solid chemical substance formed through biogeochemical processes, having characteristic chemical composition, highly ordered atomic structure, and specific physical properties. By comparison, a rock is an aggregate of minerals and/or mineraloids and does not have a specific chemical composition. Minerals range in composition from pure elements and simple salts to very complex silicates with thousands of known forms.
Rock cycle	The rock cycle is a fundamental concept in geology that describes the dynamic transitions through geologic time among the three main rock types: sedimentary, metamorphic, and igneous. As the diagram to the right illustrates, each type of rock is altered or destroyed when it is forced out of its equilibrium conditions. An igneous rock such as basalt may break down and dissolve when exposed to the atmosphere, or melt as it is subducted under a continent.
Global warming	Global warming refers to the rising average temperature of Earth's atmosphere and oceans, which began to increase in the late 19th century and is projected to continue rising. Since the early 20th century, Earth's average surface temperature has increased by about 0.8 °C (1.4 °F), with about two thirds of the increase occurring since 1980. Warming of the climate system is unequivocal, and scientists are more than 90% certain that most of it is caused by increasing concentrations of greenhouse gases produced by human activities such as deforestation and the burning of fossil fuels. These findings are recognized by the national science academies of all major industrialized nations.[A]

Climate model projections are summarized in the 2007 Fourth Assessment Report (AR4) by the Intergovernmental Panel on Climate Change (IPCC).

Anthracite

Anthracite is a hard, compact variety of mineral coal that has a high luster. It has the highest carbon count and contains the fewest impurities of all coals, and has the highest calorific content as compared to other types of coals such as bituminous coal and lignite.

Anthracite is the most metamorphosed type of coal (but still represents low-grade metamorphism), in which the carbon content is between 92.1% and 98%. The term is applied to those varieties of coal which do not give off tarry or other hydrocarbon vapours when heated below their point of ignition. Anthracite ignites with difficulty and burns with a short, blue, and smokeless flame.

Basalt

Basalt is a common extrusive volcanic rock. It is usually grey to black and fine-grained due to rapid cooling of lava at the surface of a planet. It may be porphyritic containing larger crystals in a fine matrix, or vesicular, or frothy scoria.

Extrusive

Extrusive refers to the mode of igneous volcanic rock formation in which hot magma from inside the Earth flows out (extrudes) onto the surface as lava or explodes violently into the atmosphere to fall back as pyroclastics or tuff. This is opposed to intrusive rock formation, in which magma does not reach the surface.

The main effect of extrusion is that the magma can cool much more quickly in the open air or under seawater, and there is little time for the growth of crystals.

Granite

Granite is a common widely occurring type of intrusive, felsic, igneous rock. Granite usually has a medium- to coarse-grained texture. Occasionally some individual crystals (phenocrysts) are larger than the groundmass, in which case the texture is known as porphyritic.

Chapter 8. Earth's Resources: Geologic Processes, Soil, and Minerals

Igneous rock	Igneous rock is one of the three main rock types, the others being sedimentary and metamorphic rock. Igneous rock is formed through the cooling and solidification of magma or lava. Igneous rock may form with or without crystallization, either below the surface as intrusive (plutonic) rocks or on the surface as extrusive (volcanic) rocks.
Obsidian	Obsidian is a naturally occurring volcanic glass formed as an extrusive igneous rock.
	It is produced when felsic lava extruded from a volcano cools rapidly with minimum crystal growth. Obsidian is commonly found within the margins of rhyolitic lava flows known as obsidian flows, where the chemical composition (high silica content) induces a high viscosity and polymerization degree of the lava.
Sediment	Sediment is naturally-occurring material that is broken down by processes of weathering and erosion, and is subsequently transported by the action of fluids such as wind, water, or ice, and/or by the force of gravity acting on the particle itself.
	Sediments are most often transported by water (fluvial processes) transported by wind (aeolian processes) and glaciers. Beach sands and river channel deposits are examples of fluvial transport and deposition, though sediment also often settles out of slow-moving or standing water in lakes and oceans.
Sedimentary rock	Sedimentary rock is a type of rock that is formed by sedimentation of material at the Earth's surface and within bodies of water. Sedimentation is the collective name for processes that cause mineral and/or organic particles (detritus) to settle and accumulate or minerals to precipitate from a solution. Particles that form a sedimentary rock by accumulating are called sediment.
Pollution	Pollution is the introduction of contaminants into a natural environment that causes instability, disorder, harm or discomfort to the ecosystem i.e. physical systems or living organisms. Pollution can take the form of chemical substances or energy, such as noise, heat or light. Pollutants, the components of pollution, can be either foreign substances/energies or naturally occurring contaminants.

Sandstone	Sandstone is a clastic sedimentary rock composed mainly of sand-sized minerals or rock grains.
	Most sandstone is composed of quartz and/or feldspar because these are the most common minerals in the Earth's crust. Like sand, sandstone may be any colour, but the most common colours are tan, brown, yellow, red, gray, pink, white and black.
Weathering	Weathering is the breaking down of rocks, soils and minerals as well as artificial materials through contact with the Earth's atmosphere, biota and waters. Weathering occurs in situ, or 'with no movement', and thus should not be confused with erosion, which involves the movement of rocks and minerals by agents such as water, ice, wind, and gravity.
Feldspar	Feldspars ($KAlSi_3O_8$ - $NaAlSi_3O_8$ - $CaAl_2Si_2O_8$) are a group of rock-forming tectosilicate minerals which make up as much as 60% of the Earth's crust.
	Feldspars crystallize from magma in both intrusive and extrusive igneous rocks, as veins, and are also present in many types of metamorphic rock. Feldspars are also found in many types of sedimentary rock.
Air pollution	Air pollution is the introduction of chemicals, particulate matter, or biological materials that cause harm or discomfort to humans or other living organisms, or cause damage to the natural environment or built environment, into the atmosphere.
	The atmosphere is a complex dynamic natural gaseous system that is essential to support life on planet Earth. Stratospheric ozone depletion due to air pollution has long been recognized as a threat to human health as well as to the Earth's ecosystems.
Primary	A primary (or gravitational primary) is the main physical body of a gravitationally bound, multi-object system. This body contributes most of the mass of that system and will generally be located near its center of mass.

	In the solar system, the Sun is the primary for all objects that orbit around it.
Ecosystem services	Humankind benefits from a multitude of resources and processes that are supplied by natural ecosystems. Collectively, these benefits are known as ecosystem services and include products like clean drinking water and processes such as the decomposition of wastes. While scientists and environmentalists have discussed ecosystem services for decades, these services were popularized and their definitions formalized by the United Nations 2004 Millennium Ecosystem Assessment (MA), a four-year study involving more than 1,300 scientists worldwide.
Erosion	Erosion is the process by which materials are removed from the surface and transported to another location. It works by hydraulic or aeolian actions and transport of solids (sediment, soil, rock and other particles) in the natural environment, and leads to the deposition of these materials elsewhere. It usually occurs due to transport by wind, water, or ice; by down-slope creep of soil and other material under the force of gravity; or by living organisms, such as burrowing animals, in the case of bioerosion.
Parent material	In soil science, parent material is the underlying geological material (generally bedrock or a superficial or drift deposit) in which soil horizons form. Soils typically inherit a great deal of structure and minerals from their parent material, and, as such, are often classified based upon their contents of consolidated or unconsolidated mineral material that has undergone some degree of physical or chemical weathering and the mode by which the materials were most recently transported.

Consolidated

Parent materials that are predominately composed of consolidated rock are termed residual parent material. |
| Topography | Topography is a field of planetary science comprising the study of surface shape and features of the Earth and other observable astronomical objects including planets, moons, and asteroids. It is also the description of such surface shapes and features (especially their depiction in maps). |

	The topography of an area can also mean the surface shape and features themselves.
Topsoil	Topsoil is the upper, outermost layer of soil, usually the top 2 inches (5.1 cm) to 8 inches (20 cm). It has the highest concentration of organic matter and microorganisms and is where most of the Earth's biological soil activity occurs. Importance Plants generally concentrate their roots in and obtain most of their nutrients from this layer.
Soil texture	Soil texture is a qualitative classification tool used in both the field and laboratory to determine classes for agricultural soils based on their physical texture. The classes are diinguished in the field by the 'textural feel' which can be further clarified by separating the relative proportions of sand, silt and clay using grading sieves: The Particle Size Diribution (PSD). The class is then used to determine crop suitability and to approximate the soils responses to environmental and management conditions such as drought or calcium (lime) requirements.
Green Revolution	Green Revolution refers to a series of research, development, and technology transfer initiatives, occurring between the 1940s and the late 1970s, that increased agriculture production around the world, beginning most markedly in the late 1960s. The initiatives, led by Norman Borlaug, the 'Father of the Green Revolution' credited with saving over a billion people from starvation, involved the development of high-yielding varieties of cereal grains, expansion of irrigation infrastructure, modernization of management techniques, distribution of hybridized seeds, synthetic fertilizers, and pesticides to farmers. The term 'Green Revolution' was first used in 1968 by former United States Agency for International Development (USAID) director William Gaud, who noted the spread of the new technologies and said,

---

These and other developments in the field of agriculture contain the makings of a new revolution.

| Porosity | Porosity is a measure of the void (i.e., 'empty') spaces in a material, and is a fraction of the volume of voids over the total volume, between 0-1, or as a percentage between 0-100%. The term is used in multiple fields including pharmaceutics, ceramics, metallurgy, materials, manufacturing, earth sciences and construction.

Void fraction in two-phase flow
In gas-liquid two-phase flow, the void fraction is defined as the fraction of the flow-channel volume that is occupied by the gas phase or, alternatively, as the fraction of the cross-sectional area of the channel that is occupied by the gas phase. |

| Cation exchange capacity | In soil science, cation exchange capacity is the capacity of a soil for ion exchange of cations between the soil and the soil solution. Cation exchange capacity is used as a measure of fertility, nutrient retention capacity, and the capacity to protect groundwater from cation contamination. The Base Cation Saturation Ratio (BCSR) is a method of interpreting soil test results that is widely used in sustainable agriculture, supported by the National Sustainable Agriculture Information Service (ATTRA) and claimed to be successfully in use on over a million acres of farmland worldwide. |

| Landfill | A landfill site (also known as tip, dump or rubbish dump and historically as a midden) is a site for the disposal of waste materials by burial and is the oldest form of waste treatment. Historically, landfills have been the most common methods of organized waste disposal and remain so in many places around the world.

Landfills may include internal waste disposal sites (where a producer of waste carries out their own waste disposal at the place of production) as well as sites used by many producers. |

Bauxite	Bauxite is an aluminium ore and is the main source of aluminium. This form of rock consists mostly of the minerals gibbsite $Al(OH)_3$, boehmite $\gamma\text{-}AlO(OH)$, and diaspore $\alpha\text{-}AlO(OH)$, in a mixture with the two iron oxides goethite and hematite, the clay mineral kaolinite, and small amounts of anatase TiO_2. he village Les Baux in southern France, where it was first recognised as containing aluminium and named by the French geologist Pierre Berthier in 1821.
Natural resource	Natural resources occur naturally within environments that exist relatively undisturbed by mankind, in a natural form. A natural resource is often characterized by amounts of biodiversity and geodiversity existent in various ecosystems. Natural resources are derived from the environment.
Mount Pinatubo	Mount Pinatubo is an active stratovolcano located on the island of Luzon, at the intersection of the borders of the Philippine provinces of Zambales, Tarlac, and Pampanga. It is located in the Tri-Cabusilan Mountain range separating the west coast of Luzon from the central plains, and is 42 km (26 mi) west of the dormant and more prominent Mount Arayat, occasionally mistaken for Pinatubo. Ancestral Pinatubo was a stratovolcano made of andesite and dacite.
Placer mining	Placer mining is the mining of alluvial deposits for minerals. This may be done by open-pit (also called open-cast mining) or by various surface excavating equipment or tunneling equipment.' It refers to mining the precious metal deposits (particularly gold and gemstones) found in alluvial deposits--deposits of sand and gravel in modern or ancient stream beds. The metal or gemstones, having been moved by stream flow from an original source such as a vein, is typically only a minuscule portion of the total deposit.
Surface mining	Surface mining is a type of mining in which soil and rock overlying the mineral deposit (the overburden) are removed. It is the opposite of underground mining, in which the overlying rock is left in place, and the mineral removed through shafts or tunnels. Surface mining began in the mid-sixteenth century and is practiced throughout the world, although the majority of surface mining occurs in North America. It gained popularity throughout the 20th century, and is now the predominant form of mining in coal beds such as those in Appalachia and America's Midwest.

Taiga	Taiga, is a biome characterized by coniferous forests.
	Taiga is the world's largest terrestrial biome. In North America it covers most of inland Canada and Alaska as well as parts of the extreme northern continental United States and is known as the Northwoods.
Sustainability	Sustainability is the capacity to endure. For humans, sustainability is the long-term maintenance of responsibility, which has environmental, economic, and social dimensions, and encompasses the concept of stewardship, the responsible management of resource use. In ecology, sustainability describes how biological systems remain diverse and productive over time, a necessary precondition for human well-being.

---

1. _____ is a field of study within geology concerned generally with the structures within the lithosphere of the Earth (or other planets) and particularly with the forces and movements that have operated in a region to create these structures.

   _____ is concerned with the orogenies and tectonic development of cratons and tectonic terranes as well as the earthquake and volcanic belts which directly affect much of the global population. Tectonic studies are also important for understanding erosion patterns in geomorphology and as guides for the economic geologist searching for petroleum and metallic ores.

   a. Tectonics
   b. Morphotectonics
   c. Neotectonics
   d. Strike-slip tectonics

2. The _____ of a biological species in an environment is the maximum population size of the species that the environment can sustain indefinitely, given the food, habitat, water and other necessities available in the environment. In population biology, _____ is defined as the environment's maximal load, which is different from the concept of population equilibrium.

   For the human population, more complex variables such as sanitation and medical care are sometimes considered as part of the necessary establishment.

   a. Juglone
   b. Space weather
   c. Space weathering
   d. Carrying capacity

3. The places known as _____s or hot spots in geology are volcanic regions thought to be fed by underlying mantle that is anomalously hot compared with the mantle elsewhere. They may be on, near to, or far from tectonic plate boundaries. There are two hypotheses to explain them.

---

a. Juglone
b. Hotspot
c. Giebichenstein
d. Global Boundary Stratotype Section and Point

4. _____s occur naturally within environments that exist relatively undisturbed by mankind, in a natural form.
A _____ is often characterized by amounts of biodiversity and geodiversity existent in various ecosystems.

_____s are derived from the environment.

a. Juglone
b. Singlet oxygen
c. Smoluchowski factor
d. Natural resource

5. The _____ consists of the Sun and the astronomical objects gravitationally bound in orbit around it, all of which formed from the collapse of a giant molecular cloud approximately 4.6 billion years ago. The vast majority of the system's ma (well over 99%) is in the Sun. Of the many objects that orbit the Sun, most of the ma is contained within eight relatively solitary planets[e] whose orbits are almost circular and lie within a nearly flat disc called the ecliptic plane.

a. Solar transit
b. Solar System
c. Space weathering
d. Spacewatch

---

1. a
2. d
3. b
4. d
5. b

You can take the complete Chapter Practice Test

for Chapter 8. Earth`s Resources: Geologic Processes, Soil, and Minerals
on all key terms, persons, places, and concepts.

Online 99 Cents

http://www.epub89.16.20190.8.cram101.com/

Use www.Cram101.com for all your study needs

including Cram101's online interactive problem solving labs in chemistry, statistics, mathematics, and more.

_____ | Aquifer

_____ | Groundwater

_____ | Groundwater recharge

_____ | Saltwater intrusion

_____ | Surface water

_____ | Intrusion

_____ | Atmosphere

_____ | Floodplain

_____ | Wetland

_____ | Convection

_____ | Ecosystem services

_____ | Drought

_____ | Dust Bowl

_____ | Air pollution

_____ | Pollution

_____ | Engineering

_____ | Fish ladder

_____ | Three Gorges Dam

_____ | Colorado River Aqueduct

	Distillation
	Reverse osmosis
	Green Revolution
	Drip irrigation
	Nuclear power
	Rights
	Washing
	Washing machine
	Sustainability
	Natural resource
	Water right

CHAPTER HIGHLIGHTS: KEY TERMS, PEOPLE, PLACES, CONCEPTS
Chapter 9. Water Resources: Supply, Distribution, and Use

139

Aquifer	An aquifer is a wet underground layer of water-bearing permeable rock or unconsolidated materials (gravel, sand, or silt) from which groundwater can be usefully extracted using a water well. The study of water flow in aquifers and the characterization of aquifers is called hydrogeology. Related terms include aquitard, which is a bed of low permeability along an aquifer, and aquiclude (or aquifuge), which is a solid, impermeable area underlying or overlying an aquifer.
Groundwater	Groundwater is water located beneath the ground surface in soil pore spaces and in the fractures of rock formations. A unit of rock or an unconsolidated deposit is called an aquifer when it can yield a usable quantity of water. The depth at which soil pore spaces or fractures and voids in rock become completely saturated with water is called the water table.
Groundwater recharge	Groundwater recharge is a hydrologic process where water moves downward from surface water to groundwater. This process usually occurs in the vadose zone below plant roots and is often expressed as a flux to the water table surface. Recharge occurs both naturally (through the water cycle) and anthropologically (i.e., 'artificial groundwater recharge'), where rainwater and or reclaimed water is routed to the subsurface.
Saltwater intrusion	Saltwater intrusion is the movement of saline water into freshwater aquifers. Most often, it is caused by ground-water pumping from coastal wells, or from construction of navigation channels or oil field canals. The channels and canals provide conduits for salt water to be brought into fresh water marshes.
Surface water	Surface water is water collecting on the ground or in a stream, river, lake, wetland, or ocean; it is related to water collecting as groundwater or atmospheric water.

Surface water is naturally replenished by precipitation and naturally lost through discharge to evaporation and sub-surface seepage into the ground. Although there are other sources of groundwater, such as connate water and magmatic water, precipitation is the major one and groundwater originated in this way is called meteoric water. |
| Intrusion | An intrusion is liquid rock that forms under the surface of the earth. Magma from under the surface is slowly pushed up from deep within the earth into any cracks or spaces it can find, sometimes pushing existing country rock out of the way, a process that can take millions of years. As the rock slowly cools into a solid, the different parts of the magma crystallize into minerals. |

Atmosphere	The standard atmosphere (symbol: atm) is an international reference pressure defined as 101325 Pa and formerly used as unit of pressure. For practical purposes it has been replaced by the bar which is 10^5 Pa. The difference of about 1% is not significant for many applications, and is within the error range of common pressure gges.
Floodplain	A floodplain, is a flat or nearly flat land adjacent a stream or river that stretches from the banks of its channel to the base of the enclosing valley walls and experiences flooding during periods of high discharge. It includes the floodway, which consists of the stream channel and adjacent areas that carry flood flows, and the flood fringe, which are areas covered by the flood, but which do not experience a strong current. In other words, a floodplain is an area near a river or a stream which floods easily.
Wetland	A wetland is a land area that is saturated with water, either permanently or seasonally, such that it takes on characteristics that distinguish it as a distinct ecosystem. The primary factor that distinguishes wetlands is the characteristic vegetation that is adapted to its unique soil conditions: Wetlands are made up primarily of hydric soil, which supports aquatic plants. The water found in wetlands can be saltwater, freshwater, or brackish.
Convection	Convection is the concerted, collective movement of ensembles of molecules within fluids (i.e. liquids, gases) and rheids. Convection of mass cannot take place in solids, since neither bulk current flows nor significant diffusion can take place in solids. Diffusion of heat can take place in solids, but is referred to separately in that case as heat conduction.
Ecosystem services	Humankind benefits from a multitude of resources and processes that are supplied by natural ecosystems. Collectively, these benefits are known as ecosystem services and include products like clean drinking water and processes such as the decomposition of wastes. While scientists and environmentalists have discussed ecosystem services for decades, these services were popularized and their definitions formalized by the United Nations 2004 Millennium Ecosystem Assessment (MA), a four-year study involving more than 1,300 scientists worldwide.
Drought	A drought is an extended period of months or years when a region notes a deficiency in its water supply whether surface or underground water. Generally, this occurs when a region receives consistently below average precipitation. It can have a substantial impact on the ecosystem and agriculture of the affected region.

Dust Bowl	The Dust Bowl, was a period of severe dust storms causing major ecological and agricultural damage to American and Canadian prairie lands from 1930 to 1936 (in some areas until 1940). The phenomenon was caused by severe drought coupled with decades of extensive farming without crop rotation, fallow fields, cover crops or other techniques to prevent wind erosion. Deep plowing of the virgin topsoil of the Great Plains had displaced the natural deep-rooted grasses that normally kept the soil in place and trapped moisture even during periods of drought and high winds.
Air pollution	Air pollution is the introduction of chemicals, particulate matter, or biological materials that cause harm or discomfort to humans or other living organisms, or cause damage to the natural environment or built environment, into the atmosphere. The atmosphere is a complex dynamic natural gaseous system that is essential to support life on planet Earth. Stratospheric ozone depletion due to air pollution has long been recognized as a threat to human health as well as to the Earth's ecosystems.
Pollution	Pollution is the introduction of contaminants into a natural environment that causes instability, disorder, harm or discomfort to the ecosystem i.e. physical systems or living organisms. Pollution can take the form of chemical substances or energy, such as noise, heat or light. Pollutants, the components of pollution, can be either foreign substances/energies or naturally occurring contaminants.
Engineering	Engineering is the discipline, art, skill, profession, and technology of acquiring and applying scientific, mathematical, economic, social, and practical knowledge, in order to design and build structures, machines, devices, systems, materials and processes. The American Engineers' Council for Professional Development (ECPD, the predecessor of ABET) has defined 'engineering' as: The creative application of scientific principles to design or develop structures, machines, apparatus, or manufacturing processes, or works utilizing them singly or in combination; or to construct or operate the same with full cognizance of their design; or to forecast their behavior under specific operating conditions; all as respects an intended function, economics of operation and safety to life and property.

	One who practices engineering is called an engineer, and those licensed to do so may have more formal designations such as Professional Engineer, Chartered Engineer, Incorporated Engineer, Ingenieur or European Engineer.
Fish ladder	A fish ladder, fish pass or fish steps, is a structure on or around artificial barriers (such as dams and locks) to facilitate diadromous fishes' natural migration. Most fishways enable fish to pass around the barriers by swimming and leaping up a series of relatively low steps (hence the term ladder) into the waters on the other side. The velocity of water falling over the steps has to be great enough to attract the fish to the ladder, but it cannot be so great that it washes fish back downstream or exhausts them to the point of inability to continue their journey upriver.
Three Gorges Dam	The Three Gorges Dam is a hydroelectric dam that spans the Yangtze River by the town of Sandouping, located in the Yiling District of Yichang, in Hubei province, China. The Three Gorges Dam is the world's largest power station in terms of installed capacity (21,000 MW) but is second to Itaipu Dam with regard to the generation of electricity annually.
	The dam body was completed in 2006. Except for a ship lift, the originally planned components of the project were completed on October 30, 2008, when the 26th turbine in the shore plant began commercial operation.
Colorado River Aqueduct	The Colorado River Aqueduct, is a 242 mi (389 km) water conveyance in Southern California in the United States, operated by the Metropolitan Water District of Southern California (MWD). The aqueduct impounds water from the Colorado River at Lake Havasu on the California-Arizona border west across the Mojave and Colorado deserts to the east side of the Santa Ana Mountains. It is one of the primary sources of drinking water for Southern California.
Distillation	Distillation is a method of separating mixtures based on differences in volatilities of components in a boiling liquid mixture. Distillation is a unit operation, or a physical separation process, and not a chemical reaction.
	Commercially, distillation has a number of applications.

Reverse osmosis	Reverse osmosis is a filtration method that removes many types of large molecules and ions from solutions by applying pressure to the solution when it is on one side of a selective membrane. The result is that the solute is retained on the pressurized side of the membrane and the pure solvent is allowed to pass to the other side. To be 'selective,' this membrane should not allow large molecules or ions through the pores (holes), but should allow smaller components of the solution (such as the solvent) to pass freely.
Green Revolution	Green Revolution refers to a series of research, development, and technology transfer initiatives, occurring between the 1940s and the late 1970s, that increased agriculture production around the world, beginning most markedly in the late 1960s.
	The initiatives, led by Norman Borlaug, the 'Father of the Green Revolution' credited with saving over a billion people from starvation, involved the development of high-yielding varieties of cereal grains, expansion of irrigation infrastructure, modernization of management techniques, distribution of hybridized seeds, synthetic fertilizers, and pesticides to farmers.
	The term 'Green Revolution' was first used in 1968 by former United States Agency for International Development (USAID) director William Gaud, who noted the spread of the new technologies and said,
	These and other developments in the field of agriculture contain the makings of a new revolution.
Drip irrigation	Drip irrigation, is an irrigation method which saves water and fertilizer by allowing water to drip slowly to the roots of plants, either onto the soil surface or directly onto the root zone, through a network of valves, pipes, tubing, and emitters.It is done with the help of narrow tubes which deliver water directly to the base of the plant.
	History Heda irrigation has been used since ancient times when buried clay pots were filled with water, which would gradually seep into the grass. Modern drip irrigation began its development in Afghanistan in 1866 when researchers began experimenting with irrigation using clay pipe to create combination irrigation and drainage systems.

Nuclear power	Nuclear power is the use of sustained nuclear fission to generate heat and electricity. Nuclear power plants provide about 6% of the world's energy and 13-14% of the world's electricity, with the U.S., France, and Japan together accounting for about 50% of nuclear generated electricity. In 2007, the IAEA reported there were 439 nuclear power reactors in operation in the world, operating in 31 countries.
Rights	Rights are legal, social, or ethical principles of freedom or entitlement; that is, rights are the fundamental normative rules about what is allowed of people or owed to people, according to some legal system, social convention, or ethical theory. Rights are often considered fundamental to civilization, being regarded as established pillars of society and culture, and the history of social conflicts can be found in the history of each right and its development. Rights are of essential importance in such disciplines as law and ethics, especially theories of justice and deontology.
Washing	Woman washing clothes by a canal in Batavia 1900-1940 Washing is one way of cleaning, namely with water and often some kind of soap or detergent. Washing is an essential part of good hygiene and health. Often people use soaps and detergents to assist in the emulsification of oils and dirt particles so they can be washed away.
Washing machine	A washing machine is a machine designed to wash laundry, such as clothing, towels and sheets. The term is mostly applied only to machines that use water as the primary cleaning solution, as opposed to dry cleaning (which uses alternative cleaning fluids, and is performed by specialist businesses) or even ultrasonic cleaners. History Laundering by hand involves beating and scrubbing dirty cloth.
Sustainability	Sustainability is the capacity to endure. For humans, sustainability is the long-term maintenance of responsibility, which has environmental, economic, and social dimensions, and encompasses the concept of stewardship, the responsible management of resource use. In ecology, sustainability describes how biological systems remain diverse and productive over time, a necessary precondition for human well-being.

Natural resource	Natural resources occur naturally within environments that exist relatively undisturbed by mankind, in a natural form. A natural resource is often characterized by amounts of biodiversity and geodiversity existent in various ecosystems.
	Natural resources are derived from the environment.
Water right	Water right in water law refers to the right of a user to use water from a water source, e.g., a river, stream, pond or source of groundwater. In areas with plentiful water and few users, such systems are generally not complicated or contentious. In other areas, especially arid areas where irrigation is practiced, such systems are often the source of conflict, both legal and physical.

1. The standard _____(symbol: atm) is an international reference pressure defined as 101325 Pa and formerly used as unit of pressure. For practical purposes it has been replaced by the bar which is 10^5 Pa. The difference of about 1% is not significant for many applications, and is within the error range of common pressure gges.

    a. Atmospheric dispersion modeling
    b. Atmosphere
    c. Ecological sanitation
    d. EcoProIT

2. The _____, was a period of severe dust storms causing major ecological and agricultural damage to American and Canadian prairie lands from 1930 to 1936 (in some areas until 1940). The phenomenon was caused by severe drought coupled with decades of extensive farming without crop rotation, fallow fields, cover crops or other techniques to prevent wind erosion. Deep plowing of the virgin topsoil of the Great Plains had displaced the natural deep-rooted grasses that normally kept the soil in place and trapped moisture even during periods of drought and high winds.

    a. Dust Bowl
    b. Flood-meadow
    c. Floodplain restoration
    d. Fluvial terrace

3. _____ is a method of separating mixtures based on differences in volatilities of components in a boiling liquid mixture. _____ is a unit operation, or a physical separation process, and not a chemical reaction.

    Commercially, _____ has a number of applications.

    a. Drying
    b. Flocculation
    c. Fluid extract
    d. Distillation

4. An _____ is a wet underground layer of water-bearing permeable rock or unconsolidated materials (gravel, sand, or silt) from which groundwater can be usefully extracted using a water well. The study of water flow in _____s and the characterization of _____s is called hydrogeology. Related terms include aquitard, which is a bed of low permeability along an _____, and aquiclude (or aquifuge), which is a solid, impermeable area underlying or overlying an _____.

---

a. Aquifer

b. Evapotranspiration

c. Integrated constructed wetland

d. International Joint Commission

5. _____ is water located beneath the ground surface in soil pore spaces and in the fractures of rock formations. A unit of rock or an unconsolidated deposit is called an aquifer when it can yield a usable quantity of water. The depth at which soil pore spaces or fractures and voids in rock become completely saturated with water is called the water table.

a. Hard engineering

b. Hydraulic redistribution

c. Hydrological code

d. Groundwater

---

1. b
2. a
3. d
4. a
5. d

You can take the complete Chapter Practice Test

for Chapter 9. Water Resources: Supply, Distribution, and Use
on all key terms, persons, places, and concepts.

Online 99 Cents

http://www.epub89.16.20190.9.cram101.com/

Use www.Cram101.com for all your study needs

including Cram101's online interactive problem solving labs in chemistry, statistics, mathematics, and more.

Land: Public and Private

Tragedy of the commons

Carrying capacity

Habitat

Biodiversity

Biodiversity hotspot

Hotspot

Maximum sustainable yield

Sustainable yield

National park

Land management

Deforestation

Reforestation

Plantation

Air pollution

Pollution

Clean Water Act

Endangered Species Act

Environmental policy

Environmental mitigation

National Environmental Policy Act

Urban sprawl

Smart growth

Sustainable development

Sense of place

Sustainability

Tragedy of the commons	The tragedy of the commons is a dilemma arising from the situation in which multiple individuals, acting independently and rationally consulting their own self-interest, will ultimately deplete a shared limited resource, even when it is clear that it is not in anyone's long-term interest for this to happen Theories and examples Central to Hardin's article is an example (first sketched in an 1833 pamphlet by William Forster Lloyd) involving medieval land tenure in Europe, of herders sharing a common parcel of land, on which they are each entitled to let their cows graze.
Carrying capacity	The carrying capacity of a biological species in an environment is the maximum population size of the species that the environment can sustain indefinitely, given the food, habitat, water and other necessities available in the environment. In population biology, carrying capacity is defined as the environment's maximal load, which is different from the concept of population equilibrium. For the human population, more complex variables such as sanitation and medical care are sometimes considered as part of the necessary establishment.
Habitat	A habitat is an ecological or environmental area that is inhabited by a particular species of animal, plant or other type of organism. It is the natural environment in which an organism lives, or the physical environment that surrounds (influences and is utilized by) a species population. Definition The term 'population' is preferred to 'organism' because, while it is possible to describe the habitat of a single black bear, it is also possible that one may not find any particular or individual bear but the grouping of bears that constitute a breeding population and occupy a certain biogeographical area.
Biodiversity	Biodiversity is the degree of variation of life forms within a given species, ecosystem, biome, or an entire planet. Biodiversity is a measure of the health of ecosystems. Biodiversity is in part a function of climate.

Biodiversity hotspot	A biodiversity hotspot is a biogeographic region with a significant reservoir of biodiversity that is under threat from humans. The concept of biodiversity hotspots was originated by Norman Myers in two articles in 'The Environmentalist' (1988 ' 1990), revised after thorough analysis by Myers and others in 'Hotspots: Earth's Biologically Richest and Most Endangered Terrestrial Ecoregions'. To qualify as a biodiversity hotspot on Myers 2000 edition of the hotspot-map, a region must meet two strict criteria: it must contain at least 0.5% or 1,500 species of vascular plants as endemics, and it has to have lost at least 70% of its primary vegetation.
Hotspot	The places known as hotspots or hot spots in geology are volcanic regions thought to be fed by underlying mantle that is anomalously hot compared with the mantle elsewhere. They may be on, near to, or far from tectonic plate boundaries. There are two hypotheses to explain them.
Maximum sustainable yield	In population ecology and economics, maximum sustainable yield is theoretically, the largest yield (or catch) that can be taken from a species' stock over an indefinite period. Fundamental to the notion of sustainable harvest, the concept of aims to maintain the population size at the point of maximum growth rate by harvesting the individuals that would normally be added to the population, allowing the population to continue to be productive indefinitely. Under the assumption of logistic growth, resource limitation does not constrain individuals' reproductive rates when populations are small, but because there are few individuals, the overall yield is small.
Sustainable yield	The sustainable yield of natural capital is the ecological yield that can be extracted without reducing the base of capital itself, i.e. the surplus required to maintain ecostem services at the same or increasing level over time. This yield usually varies over time with the needs of the ecostem to maintain itself, e.g. a forest that has recently suffered a blight or flooding or fire will require more of its own ecological yield to sustain and re-establish a mature forest. While doing so, the sustainable yield may be much less.
National park	A national park is a reserve of natural, semi-natural, or developed land that a sovereign state declares or owns. Although individual nations designate their own national parks differently , an international organization, the International Union for Conservation of Nature (IUCN), and its World Commission on Protected Areas, has defined National Parks as its category II type of protected areas.

Chapter 10. Land: Public and Private

Land management	Land management is the process of managing the use and development (in both urban and rural settings) of land resources. Land resources are used for a variety of purposes which may include organic agriculture, reforestation, water resource management and eco-tourism projects.
Deforestation	Deforestation is the removal of a forest or stand of trees where the land is thereafter converted to a nonforest use. Examples of deforestation include conversion of forestland to farms, ranches, or urban use.

The term deforestation is often misused to describe any activity where all trees in an area are removed. |
| Reforestation | Reforestation is the natural or intentional restocking of existing forests and woodlands that have been depleted, usually through deforestation. Reforestation can be used to improve the quality of human life by soaking up pollution and dust from the air, rebuild natural habitats and ecosystems, mitigate global warming since forests facilitate biosequestration of atmospheric carbon dioxide, and harvest for resources, particularly timber.

The term reforestation is similar to afforestation, the process of restoring and recreating areas of woodlands or forests that may have existed long ago but were deforested or otherwise removed at some point in the past. |
| Plantation | A plantation is a large artificially established forest, farm or estate, where crops are grown for sale, often in distant markets rather than for local on-site consumption. The term plantation is informal and not precisely defined. |
| Air pollution | Air pollution is the introduction of chemicals, particulate matter, or biological materials that cause harm or discomfort to humans or other living organisms, or cause damage to the natural environment or built environment, into the atmosphere. |

The atmosphere is a complex dynamic natural gaseous system that is essential to support life on planet Earth. Stratospheric ozone depletion due to air pollution has long been recognized as a threat to human health as well as to the Earth's ecosystems.

Pollution	Pollution is the introduction of contaminants into a natural environment that causes instability, disorder, harm or discomfort to the ecosystem i.e. physical systems or living organisms. Pollution can take the form of chemical substances or energy, such as noise, heat or light. Pollutants, the components of pollution, can be either foreign substances/energies or naturally occurring contaminants.
Clean Water Act	The Clean Water Act is the primary federal law in the United States governing water pollution. Commonly abbreviated as the Clean Water Act, the act established the goals of eliminating releases of high amounts of toxic substances into water, eliminating additional water pollution by 1985, and ensuring that surface waters would meet standards necessary for human sports and recreation by 1983.
	The principal body of law currently in effect is based on the Federal Water Pollution Control Amendments of 1972 and was significantly expanded from the Federal Water Pollution Control Amendments of 1948. Major amendments were enacted in the Clean Water Act of 1977 and the Water Quality Act of 1987.
Endangered Species Act	The Endangered Species Act of 1973 (Endangered Species Act; 7 U.S.C. § 136, 16 U.S.C. § 1531 et seq). is one of the dozens of United States environmental laws passed in the 1970s. Signed into law by President Richard Nixon on December 28, 1973, it was designed to protect critically imperiled species from extinction as a 'consequence of economic growth and development untempered by adequate concern and conservation.'
	The Act is administered by two federal agencies, the United States Fish and Wildlife Service (FWS) and the National Oceanic and Atmospheric Administration (NOAA).

Environmental policy	Environmental policy is any [course of] action deliberately taken [or not taken] to manage human activities with a view to prevent, reduce, or mitigate harmful effects on nature and natural resources, and ensuring that man-made changes to the environment do not have harmful effects on humans. Definition It is useful to consider that environmental policy comprises two major terms: environment and policy. Environment primarily refers to the ecological dimension (ecosystems), but can also take account of social dimension (quality of life) and an economic dimension (resource management).
Environmental mitigation	Environmental mitigation, compensatory mitigation, or mitigation banking, are terms used primarily by the United States government and the related environmental industry to describe projects or programs intended to offset known impacts to an existing historic or natural resource such as a stream, wetland, endangered species, archeological site or historic structure. To 'mitigate' means to make less harsh or hostile. Environmental mitigation is typically a part of an environmental crediting syst established by governing bodies which involves allocating debits and credits.
National Environmental Policy Act	The National Environmental Policy Act is a United States environmental law that established a U.S. national policy promoting the enhancement of the environment and also established the President's Council on Environmental Quality (CEQ). Eccleston writes that as one of the most emulated statutes in the world, National Environmental Policy Act has been called the modern day equivalent of an 'Environmental Magna Carta.' National Environmental Policy Act's most significant effect was to set up procedural requirements for all federal government agencies to prepare Environmental Assessments (EAs) and Environmental Impact Statements (EISs). EAs and EISs contain statements of the environmental effects of proposed federal agency actions.
Urban sprawl	Urban sprawl, is a multifaceted concept, which includes the spreading outwards of a city and its suburbs to its outskirts to low-density and auto-dependent development on rural land, high segregation of es (e.g. stores and residential), and vario design features that encourage car dependency.

Discsions and debates about sprawl are often obfcated by the ambiguity associated with the phrase. For example, some commentators measure sprawl only with the average number of residential units per acre in a given area.

Smart growth	Smart growth is an urban planning and transportation theory that concentrates growth in compact walkable urban centers to avoid sprawl and advocates compact, transit-oriented, walkable, bicycle-friendly land use, including neighborhood schools, complete streets, and mixed-use development with a range of housing choices. The term 'smart growth' is particularly used in North America. In Europe and particularly the UK, the terms 'Compact City' or 'urban intensification' have often been used to describe similar concepts, which have influenced Government planning policies in the UK, the Netherlands and several other European countries.
Sustainable development	Sustainable development is a pattern of growth in which resource use aims to meet human needs while preserving the environment so that these needs can be met not only in the present, but also for generations to come (sometimes taught as ELF-Environment, Local people, Future). The term sustainable development was used by the Brundtland Commission which coined what has become the most often-quoted definition of sustainable development as development that 'meets the needs of the present without compromising the ability of future generations to meet their own needs.'
	Sustainable development ties together concern for the carrying capacity of natural systems with the social challenges facing humanity. As early as the 1970s 'sustainability' was employed to describe an economy 'in equilibrium with basic ecological support systems.' Ecologists have pointed to The Limits to Growth, and presented the alternative of a 'steady state economy' in order to address environmental concerns.
Sense of place	The term sense of place has been defined and used in many different ways by many different people. To some, it is a characteristic that some geographic places have and some do not, while to others it is a feeling or perception held by people (not by the place itself). It is often used in relation to those characteristics that make a place special or unique, as well as to those that foster a sense of authentic human attachment and belonging.

Sustainability	Sustainability is the capacity to endure. For humans, sustainability is the long-term maintenance of responsibility, which has environmental, economic, and social dimensions, and encompasses the concept of stewardship, the responsible management of resource use. In ecology, sustainability describes how biological systems remain diverse and productive over time, a necessary precondition for human well-being.

1. _____ is the natural or intentional restocking of existing forests and woodlands that have been depleted, usually through deforestation. _____ can be used to improve the quality of human life by soaking up pollution and dust from the air, rebuild natural habitats and ecosystems, mitigate global warming since forests facilitate biosequestration of atmospheric carbon dioxide, and harvest for resources, particularly timber.

   The term _____ is similar to afforestation, the process of restoring and recreating areas of woodlands or forests that may have existed long ago but were deforested or otherwise removed at some point in the past.

   a. Reforestation
   b. K5 Plan
   c. Paperless office
   d. Rainforest Shmainforest

2. The _____ of a biological species in an environment is the maximum population size of the species that the environment can sustain indefinitely, given the food, habitat, water and other necessities available in the environment. In population biology, _____ is defined as the environment's maximal load, which is different from the concept of population equilibrium.

   For the human population, more complex variables such as sanitation and medical care are sometimes considered as part of the necessary establishment.

   a. Carrying capacity
   b. Tyranny of small decisions
   c. Water stress
   d. White certificates

3. The _____ is a dilemma arising from the situation in which multiple individuals, acting independently and rationally consulting their own self-interest, will ultimately deplete a shared limited resource, even when it is clear that it is not in anyone's long-term interest for this to happen

   Theories and examples

Central to Hardin's article is an example (first sketched in an 1833 pamphlet by William Forster Lloyd) involving medieval land tenure in Europe, of herders sharing a common parcel of land, on which they are each entitled to let their cows graze.

a. Travel cost analysis
b. Tragedy of the commons
c. Water stress
d. White certificates

4. _____ is the capacity to endure. For humans, _____ is the long-term maintenance of responsibility, which has environmental, economic, and social dimensions, and encompasses the concept of stewardship, the responsible management of resource use. In ecology, _____ describes how biological systems remain diverse and productive over time, a necessary precondition for human well-being.

a. Back-to-the-land movement
b. BEST Education Network
c. Sustainability
d. Biocapacity

5. A _____ is an ecological or environmental area that is inhabited by a particular species of animal, plant or other type of organism. It is the natural environment in which an organism lives, or the physical environment that surrounds (influences and is utilized by) a species population.

Definition
The term 'population' is preferred to 'organism' because, while it is possible to describe the _____ of a single black bear, it is also possible that one may not find any particular or individual bear but the grouping of bears that constitute a breeding population and occupy a certain biogeographical area.

a. Habitat Conservation Plan
b. Habitat destruction
c. Landscape limnology
d. Habitat

---

1. a
2. a
3. b
4. c
5. d

You can take the complete Chapter Practice Test

for Chapter 10. Land: Public and Private
on all key terms, persons, places, and concepts.

Online 99 Cents

http://www.epub89.16.20190.10.cram101.com/

Use www.Cram101.com for all your study needs

including Cram101's online interactive problem solving labs in chemistry, statistics, mathematics, and more.

_____ | Green Revolution _____

_____ | Fly ash _____

_____ | Food security _____

_____ | Overnutrition _____

_____ | Industrial agriculture _____

_____ | Aquifer _____

_____ | Bioaccumulation _____

_____ | Herbicide _____

_____ | Insecticide _____

_____ | Pesticide _____

_____ | Pollution _____

_____ | Topsoil _____

_____ | Water pollution _____

_____ | Air pollution _____

_____ | Engineering _____

_____ | Biodiversity _____

_____ | Invertebrate _____

_____ | Genetically modified organism

_____ | Slash-and-burn _____

Chapter 11. Agriculture: Feeding the World

_____ Crop rotation

_____ Desertification

_____ Intercropping

_____ Sustainable agriculture

_____ Rotation

_____ Agroforestry

_____ Contour plowing

_____ Integrated pest management

_____ Feedlot

_____ Aquaculture

_____ Tragedy of the commons

_____ Bycatch

_____ Sustainability

| Green Revolution | Green Revolution refers to a series of research, development, and technology transfer initiatives, occurring between the 1940s and the late 1970s, that increased agriculture production around the world, beginning most markedly in the late 1960s.

The initiatives, led by Norman Borlaug, the 'Father of the Green Revolution' credited with saving over a billion people from starvation, involved the development of high-yielding varieties of cereal grains, expansion of irrigation infrastructure, modernization of management techniques, distribution of hybridized seeds, synthetic fertilizers, and pesticides to farmers.

The term 'Green Revolution' was first used in 1968 by former United States Agency for International Development (USAID) director William Gaud, who noted the spread of the new technologies and said,

These and other developments in the field of agriculture contain the makings of a new revolution. |
| --- | --- |
| Fly ash | Fly ash is one of the residues generated in combustion, and comprises the fine particles that rise with the flue gases. Ash which does not rise is termed bottom ash. In an industrial context, fly ash usually refers to ash produced during combustion of coal. |
| Food security | Food security refers to the availability of food and one's access to it. A household is considered food-secure when its occupants do not live in hunger or fear of starvation. According to the World Resources Institute, global per capita food production has been increasing substantially for the past several decades. |
| Overnutrition | Overnutrition is a form of malnutrition in which nutrients are oversupplied relative to the amounts required for normal growth, development, and metabolism. Overnutrition is a type of malnutrition where there are more nutrients than required for normal growth. |

The term can refer to:

- obesity, brought on by general overeating of foods high in caloric content, as well as:
- the oversupply of a specific nutrient or categories of nutrients, such as mineral or vitamin poisoning, due to excessive intake of dietary supplements or foods high in nutrients (such as liver), or nutritional imbalances caused by various fad diets.

For mineral excess, see:

- Iron poisoning, and
- low sodium diet (excess sodium).

.

Industrial agriculture	Industrial farming is a form of modern farming that refers to the industrialized production of livestock, poultry, fish, and crops. The methods of industrial agriculture are technoscientific, economic, and political. They include innovation in agricultural machinery and farming methods, genetic technology, techniques for achieving economies of scale in production, the creation of new markets for consumption, the application of patent protection to genetic information, and global trade.
Aquifer	An aquifer is a wet underground layer of water-bearing permeable rock or unconsolidated materials (gravel, sand, or silt) from which groundwater can be usefully extracted using a water well. The study of water flow in aquifers and the characterization of aquifers is called hydrogeology. Related terms include aquitard, which is a bed of low permeability along an aquifer, and aquiclude (or aquifuge), which is a solid, impermeable area underlying or overlying an aquifer.
Bioaccumulation	Bioaccumulation refers to the accumulation of substances, such as pesticides, or other organic chemicals in an organism. Bioaccumulation occurs when an organism absorbs a toxic substance at a rate greater than that at which the substance is lost. Thus, the longer the biological half-life of the substance the greater the risk of chronic poisoning, even if environmental levels of the toxin are not very high.
Herbicide	Herbicides, also commonly known as weedkillers, are pesticides used to kill unwanted plants. Selective herbicides kill specific targets while leaving the desired crop relatively unharmed. Some of these act by interfering with the growth of the weed and are often synthetic 'imitations' of plant hormones.

Insecticide	An insecticide is a pesticide used against insects. They include ovicides and larvicides used against the eggs and larvae of insects respectively. Insecticides are used in agriculture, medicine, industry and the household.
Pesticide	Pesticides are substances or mixture of substances intended for preventing, destroying, repelling or mitigating any pest. A pesticide may be a chemical, biological agent (such as a virus or bacterium), antimicrobial, disinfectant or device used against any pest. Pests include insects, plant pathogens, weeds, molluscs, birds, mammals, fish, nematodes (roundworms), and microbes that destroy property, spread disease or are vectors for disease or cause nuisance.
Pollution	Pollution is the introduction of contaminants into a natural environment that causes instability, disorder, harm or discomfort to the ecosystem i.e. physical systems or living organisms. Pollution can take the form of chemical substances or energy, such as noise, heat or light. Pollutants, the components of pollution, can be either foreign substances/energies or naturally occurring contaminants.
Topsoil	Topsoil is the upper, outermost layer of soil, usually the top 2 inches (5.1 cm) to 8 inches (20 cm). It has the highest concentration of organic matter and microorganisms and is where most of the Earth's biological soil activity occurs. Importance Plants generally concentrate their roots in and obtain most of their nutrients from this layer.
Water pollution	Water pollution is the contamination of water bodies (e.g. lakes, rivers, oceans, aquifers and groundwater). Water pollution occurs when pollutants are discharged directly or indirectly into water bodies without adequate treatment to remove harmful compounds. Water pollution affects plants and organisms living in these bodies of water.

Chapter 11. Agriculture: Feeding the World

Air pollution	Air pollution is the introduction of chemicals, particulate matter, or biological materials that cause harm or discomfort to humans or other living organisms, or cause damage to the natural environment or built environment, into the atmosphere.
	The atmosphere is a complex dynamic natural gaseous system that is essential to support life on planet Earth. Stratospheric ozone depletion due to air pollution has long been recognized as a threat to human health as well as to the Earth's ecosystems.
Engineering	Engineering is the discipline, art, skill, profession, and technology of acquiring and applying scientific, mathematical, economic, social, and practical knowledge, in order to design and build structures, machines, devices, systems, materials and processes.
	The American Engineers' Council for Professional Development (ECPD, the predecessor of ABET) has defined 'engineering' as:
	The creative application of scientific principles to design or develop structures, machines, apparatus, or manufacturing processes, or works utilizing them singly or in combination; or to construct or operate the same with full cognizance of their design; or to forecast their behavior under specific operating conditions; all as respects an intended function, economics of operation and safety to life and property.
	One who practices engineering is called an engineer, and those licensed to do so may have more formal designations such as Professional Engineer, Chartered Engineer, Incorporated Engineer, Ingenieur or European Engineer.
Biodiversity	Biodiversity is the degree of variation of life forms within a given species, ecosystem, biome, or an entire planet. Biodiversity is a measure of the health of ecosystems. Biodiversity is in part a function of climate.

Invertebrate	An invertebrate is an animal without a backbone. The group includes 97% of all animal species - all animals except those in the chordate subphylum Vertebrata (fish, amphibians, reptiles, birds, and mammals).
	Invertebrates form a paraphyletic group.
Genetically modified organism	A genetically modified organism or genetically engineered organism (GEO) is an organism whose genetic material has been altered using genetic engineering techniques. These techniques, generally known as recombinant DNA technology, use DNA molecules from different sources, which are combined into one molecule to create a new set of genes. This DNA is then transferred into an organism, giving it modified or novel genes.
Slash-and-burn	Slash-and-burn is an agricultural technique which involves cutting and burning of forests or woodlands to create fields. It is subsistence agriculture that typically uses little technology or other tools. It is typically part of shifting cultivation agriculture, and of transhumance livestock herding.
Crop rotation	Crop rotation is the practice of growing a series of dissimilar types of crops in the same area in sequential seasons.
	Crop rotation confers various benefits to the soil. A traditional element of crop rotation is the replenishment of nitrogen through the use of green manure in sequence with cereals and other crops.
Desertification	Desertification is the degradation of land in any drylands. Caused by a variety of factors, such as climate change and human activities, desertification is one of the most significant global environmental problems.
	Definitions Considerable controversy exists over the proper definition of the term 'desertification' for which Helmut Geist (2005) has identified more than 100 formal definitions.

Chapter 11. Agriculture: Feeding the World

Intercropping	Intercropping is the practice of growing two or more crops in proximity. The most common goal of intercropping is to produce a greater yield on a given piece of land by making use of resources that would otherwise not be utilized by a single crop. Careful planning is required, taking into account the soil, climate, crops, and varieties.
Sustainable agriculture	Sustainable agriculture is the practice of farming using principles of ecology, the study of relationships between organisms and their environment. It has been defined as 'an integrated system of plant and animal production practices having a site-specific application that will last over the long term: • tisfy human food and fiber needs • Enhance environmental quality and the natural resource base upon which the agricultural economy depends • Make the most efficient use of non-renewable resources and on-farm resources and integrate, where appropriate, natural biological cycles and controls • Sustain the economic viability of farm operations • Enhance the quality of life for farmers and society as a whole.' Sustainable agriculture in the United States was addressed by the 1990 farm bill. More recently, as consumer and retail demand for sustainable products has risen, organizations such as Food Alliance and Protected Harvest have started to provide measurement standards and certification programs for what constitutes a sustainably grown crop.
Rotation	A rotation is a circular movement of an object around a center (or point) of rotation. A three-dimensional object rotates always around an imaginary line called a rotation axis. If the axis is within the body, and passes through its center of mass the body is said to rotate upon itself, or spin.
Agroforestry	Agroforestry is an integrated approach of using the interactive benefits from combining trees and shrubs with crops and/or livestock. It combines agricultural and forestry technologies to create more diverse, productive, profitable, healthy and sustainable land-use systems. Definitions

According to the World Agroforestry Centre, Agroforestry is a collective name for land use systems and practices in which woody perennials are deliberately integrated with crops and/or animals on the same land management unit.

Contour plowing	Contour plowing or contour farming is the farming practice of plowing across a slope following its elevation contour lines. The rows form slow water run-off during rainstorms to prevent soil erosion and allow the water time to settle into the soil. In contour plowing, the ruts made by the plow run perpendicular rather than parallel to slopes, generally resulting in furrows that curve around the land and are level.
Integrated pest management	Integrated pest management is a broad based ecological approach to agricultural pest control that integrates pesticides/herbicides into a management system incorporating a range of practices for economic control of a pest. In , one attempts to prevent infestation, to observe patterns of infestation when they occur, and to intervene (without poisons) when one deems necessary. is the intelligent selection and use of pest control actions that will ensure favourable economic, ecological and sociological consequences.
Feedlot	A feedlot is a type of animal feeding operation (AFO) which is used in factory farming for finishing livestock, notably beef cattle, but also swine, horses, sheep, turkeys, chickens or ducks, prior to slaughter. Large beef feedlots are called Concentrated Animal Feeding Operations (CAFOs). They may contain thousands of animals in an array of pens.
Aquaculture	Aquaculture, is the farming of aquatic organisms such as fish, crustaceans, molluscs and aquatic plants. Aquaculture involves cultivating freshwater and saltwater populations under controlled conditions, and can be contrasted with commercial fishing, which is the harvesting of wild fish. Mariculture refers to aquaculture practised in marine environments.
Tragedy of the commons	The tragedy of the commons is a dilemma arising from the situation in which multiple individuals, acting independently and rationally consulting their own self-interest, will ultimately deplete a shared limited resource, even when it is clear that it is not in anyone's long-term interest for this to happen Theories and examples Central to Hardin's article is an example (first sketched in an 1833 pamphlet by William Forster Lloyd) involving medieval land tenure in Europe, of herders sharing a common parcel of land, on which they are each entitled to let their cows graze.

Bycatch	The term 'bycatch' is usually used for fish caught unintentionally in a fishery while intending to catch other fish. It may however also indicate untargeted catch in other forms of animal harvesting or collecting. Bycatch is of a different species, undersized individuals of the target species, or juveniles of the target species.
Sustainability	Sustainability is the capacity to endure. For humans, sustainability is the long-term maintenance of responsibility, which has environmental, economic, and social dimensions, and encompasses the concept of stewardship, the responsible management of resource use. In ecology, sustainability describes how biological systems remain diverse and productive over time, a necessary precondition for human well-being.

---

1. _____ is the upper, outermost layer of soil, usually the top 2 inches (5.1 cm) to 8 inches (20 cm). It has the highest concentration of organic matter and microorganisms and is where most of the Earth's biological soil activity occurs.

    Importance

    Plants generally concentrate their roots in and obtain most of their nutrients from this layer.

    a. Paleosol
    b. Topsoil
    c. pedalfer
    d. Pedocal

2. _____ refers to a series of research, development, and technology transfer initiatives, occurring between the 1940s and the late 1970s, that increased agriculture production around the world, beginning most markedly in the late 1960s.

    The initiatives, led by Norman Borlaug, the 'Father of the _____' credited with saving over a billion people from starvation, involved the development of high-yielding varieties of cereal grains, expansion of irrigation infrastructure, modernization of management techniques, distribution of hybridized seeds, synthetic fertilizers, and pesticides to farmers.

    The term '_____' was first used in 1968 by former United States Agency for International Development (USAID) director William Gaud, who noted the spread of the new technologies and said,

    These and other developments in the field of agriculture contain the makings of a new revolution.

    a. Juglone
    b. Green Revolution
    c. Krakatoa
    d. Refuse Act

---

3. _____s, also commonly known as weedkillers, are pesticides used to kill unwanted plants. Selective _____s kill specific targets while leaving the desired crop relatively unharmed. Some of these act by interfering with the growth of the weed and are often synthetic 'imitations' of plant hormones.

    a. Hydrocarbon
    b. Herbicide
    c. Maximum Residue Limit
    d. Methyl tert-butyl ether

4. _____ is an integrated approach of using the interactive benefits from combining trees and shrubs with crops and/or livestock. It combines agricultural and forestry technologies to create more diverse, productive, profitable, healthy and sustainable land-use systems.

Definitions
According to the World _____ Centre, _____ is a collective name for land use systems and practices in which woody perennials are deliberately integrated with crops and/or animals on the same land management unit.

    a. Odyssey
    b. Centripetal force
    c. Agroforestry
    d. Coriolis effect

5. _____ refers to the accumulation of substances, such as pesticides, or other organic chemicals in an organism. _____ occurs when an organism absorbs a toxic substance at a rate greater than that at which the substance is lost. Thus, the longer the biological half-life of the substance the greater the risk of chronic poisoning, even if environmental levels of the toxin are not very high.

    a. Bioaccumulation
    b. Biofilter
    c. Biofouling
    d. Biomagnification

---

1. b
2. b
3. b
4. c
5. a

You can take the complete Chapter Practice Test

for Chapter 11. Agriculture: Feeding the World
on all key terms, persons, places, and concepts.

Online 99 Cents

http://www.epub89.16.20190.11.cram101.com/

Use www.Cram101.com for all your study needs

including Cram101's online interactive problem solving labs in chemistry, statistics, mathematics, and more.

Nonrenewable Energy: Coal, Oil, Natural Gas, and Nuclear Fuels

_____ | Oil spill

_____ | Pollution

_____ | Fossil

_____ | Fossil fuel

_____ | Nuclear fuel

_____ | Thermodynamics

_____ | Efficiency

_____ | Deforestation

_____ | Energy carrier

_____ | Combined cycle

_____ | Nuclear power

_____ | Air pollution

_____ | Anthracite

_____ | Bituminous coal

_____ | Lignite

_____ | Sub-bituminous coal

_____ | Carbon cycle

_____ | Oil sands

_____ | Global warming

_____ | Peak oil ____

_____ | Radioactive decay ____

_____ | Water pollution ____

_____ | Radiation ____

_____ | Radioactive waste ____

_____ | Nuclear fusion ____

_____ | Sustainability ____

Oil spill	An oil spill is the release of a liquid petroleum hydrocarbon into the environment, especially marine areas, due to human activity, and is a form of pollution. The term is mtly used to describe marine oil spills, where oil is released into the ocean or coastal waters. Oil spills may be due to releases of crude oil from tankers, offshore platforms, drilling rigs and wells, as well as spills of refined petroleum products (such as gasoline, diesel) and their by-products, heavier fuels used by large ships such as bunker fuel, or the spill of any oily refuse or waste oil.
Pollution	Pollution is the introduction of contaminants into a natural environment that causes instability, disorder, harm or discomfort to the ecosystem i.e. physical systems or living organisms. Pollution can take the form of chemical substances or energy, such as noise, heat or light. Pollutants, the components of pollution, can be either foreign substances/energies or naturally occurring contaminants.
Fossil	Fossils are the preserved remains or traces of animals (also known as zoolites), plants, and other organisms from the remote past. The totality of fossils, both discovered and undiscovered, and their placement in fossiliferous (fossil-containing) rock formations and sedimentary layers (strata) is known as the fossil record.

The study of fossils across geological time, how they were formed, and the evolutionary relationships between taxa (phylogeny) are some of the most important functions of the science of paleontology. |
| Fossil fuel | Fossil fuels are fuels formed by natural processes such as anaerobic decomposition of buried dead organisms. The age of the organisms and their resulting fossil fuels is typically millions of years, and sometimes exceeds 650 million years. Fossil fuels contain high percentages of carbon and include coal, petroleum, and natural gas. |
| Nuclear fuel | Nuclear fuel is a material that can be 'consumed' by nuclear fission or fusion to derive nuclear energy. Nuclear fuel can refer to the fuel itself, or to physical objects (for example bundles composed of fuel rods) composed of the fuel material, mixed with structural, neutron moderating, or neutron reflecting materials.

Most nuclear fuels contain heavy fissile elements that are capable of nuclear fission. |

Thermodynamics	ImgProperty database img Thermodynamics is the branch of physical science concerned with heat and its relation to other forms of energy and work. It defines macroscopic variables (such as temperature, entropy, and pressure) that describe average properties of material bodies and radiation, and explains how they are related and by what laws they change with time. Thermodynamics does not describe the microscopic constituents of matter, and its laws can be derived from statistical mechanics.
Efficiency	Efficiency in general describes the extent to which time or effort is well used for the intended task or purpose. It is often used with the specific purpose of relaying the capability of a specific application of effort to produce a specific outcome effectively with a minimum amount or quantity of waste, expense, or unnecessary effort. 'Efficiency' has widely varying meanings in different disciplines.
Deforestation	Deforestation is the removal of a forest or stand of trees where the land is thereafter converted to a nonforest use. Examples of deforestation include conversion of forestland to farms, ranches, or urban use. The term deforestation is often misused to describe any activity where all trees in an area are removed.
Energy carrier	According to ISO 13600, an energy carrier is either a substance (energy form) or a phenomenon (energy system) that can be used to produce mhanical work or heat or to operate chemical or physical processes. In the field of Energetics, however, an energy carrier corresponds only to an energy form (not an energy system) of energy input required by the various stors of society to perform their functions.

Examples of energy carriers include liquid fuel in a furnace, gasoline in a pump, eltricity in a factory or a house, and hydrogen in a tank of a car.

Combined cycle	ImgProperty database img In electric power generation a combined cycle is an assembly of heat engines that work in tandem off the same source of heat, converting it into mechanical energy, which in turn usually drives electrical generators. The principle is that the exhaust of one heat engine is used as the heat source for another, thus extracting more useful energy from the heat, increasing the system's overall efficiency. This works because heat engines are only able to use a portion of the energy their fuel generates (usually less than 50%).
Nuclear power	Nuclear power is the use of sustained nuclear fission to generate heat and electricity. Nuclear power plants provide about 6% of the world's energy and 13-14% of the world's electricity, with the U.S., France, and Japan together accounting for about 50% of nuclear generated electricity. In 2007, the IAEA reported there were 439 nuclear power reactors in operation in the world, operating in 31 countries.
Air pollution	Air pollution is the introduction of chemicals, particulate matter, or biological materials that cause harm or discomfort to humans or other living organisms, or cause damage to the natural environment or built environment, into the atmosphere. The atmosphere is a complex dynamic natural gaseous system that is essential to support life on planet Earth. Stratospheric ozone depletion due to air pollution has long been recognized as a threat to human health as well as to the Earth's ecosystems.
Anthracite	Anthracite is a hard, compact variety of mineral coal that has a high luster. It has the highest carbon count and contains the fewest impurities of all coals, and has the highest calorific content as compared to other types of coals such as bituminous coal and lignite.

Anthracite is the most metamorphosed type of coal (but still represents low-grade metamorphism), in which the carbon content is between 92.1% and 98%. The term is applied to those varieties of coal which do not give off tarry or other hydrocarbon vapours when heated below their point of ignition. Anthracite ignites with difficulty and burns with a short, blue, and smokeless flame.

Bituminous coal	Bituminous coal is a relatively soft coal containing a tarlike substance called bitumen. It is of higher quality than lignite coal but of poorer quality than anthracite. It was usually formed as a result of high pressure on lignite.
Lignite	Lignite, often referred to as brown coal, or Rosebud coal by Northern Pacific Railroad, is a soft brown fuel with characteristics that put it somewhere between coal and peat. It is considered the lowest rank of coal; it is mined in Greece, Germany, Poland, Serbia, Russia, the United States, India, Australia and many other parts of Europe and it is used almost exclusively as a fuel for steam-electric power generation. Up to 50% of Greece's electricity and 24.6% of Germany's comes from lignite power plants.
Sub-bituminous coal	Sub-bituminous coal is a type of coal whose properties range from those of lignite to those of bituminous coal and are used primarily as fuel for steam-electric power generation.
	Sub-bituminous coals may be dull, dark brown to black, soft and crumbly at the lower end of the range, to bright jet-black, hard, and relatively strong at the upper end. They contain 15-30% inherent moisture by weight and are non-coking (undergo little swelling upon heating).
Carbon cycle	The carbon cycle is the biogeochemical cycle by which carbon is exchanged among the biosphere, pedosphere, geosphere, hydrosphere, and atmosphere of the Earth. It is one of the most important cycles of the earth and allows for carbon to be recycled and reused throughout the biosphere and all of its organisms.
	The carbon cycle was initially discovered by Joseph Priestley and Antoine Lavoisier, and popularized by Humphry Davy.

Oil sands	Bituminous sands, colloquially known as oil sands, are a type of unconventional petroleum deposit. The sands contain naturally occurring mixtures of sand, clay, water, and a dense and extremely viscous form of petroleum technically referred to as bitumen (or colloquially 'tar' due to its similar appearance, odour, and colour). Oil sands are found in large amounts in many countries throughout the world, but are found in extremely large quantities in Canada and Venezuela.
Global warming	Global warming refers to the rising average temperature of Earth's atmosphere and oceans, which began to increase in the late 19th century and is projected to continue rising. Since the early 20th century, Earth's average surface temperature has increased by about 0.8 °C (1.4 °F), with about two thirds of the increase occurring since 1980. Warming of the climate system is unequivocal, and scientists are more than 90% certain that most of it is caused by increasing concentrations of greenhouse gases produced by human activities such as deforestation and the burning of fossil fuels. These findings are recognized by the national science academies of all major industrialized nations.[A] Climate model projections are summarized in the 2007 Fourth Assessment Report (AR4) by the Intergovernmental Panel on Climate Change (IPCC).
Peak oil	Peak oil is the int in time when the maximum rate of global petroleum extraction is reached, after which the rate of production enters terminal decline. This concept is based on the observed production rates of individual oil wells, projected reserves and the combined production rate of a field of related oil wells. In order to understand physical Peak oil, the growing effort for production must be considered.
Radioactive decay	Radioactive decay is the process by which an atomic nucleus of an unstable atom loses energy by emitting ionizing particles (ionizing radiation) A decay, or loss of energy, results when an atom with one type of nucleus, called the parent radionuclide, transforms to an atom with a nucleus in a different state, or to a different nucleus containing different numbers of nucleons.
Water pollution	Water pollution is the contamination of water bodies (e.g. lakes, rivers, oceans, aquifers and groundwater). Water pollution occurs when pollutants are discharged directly or indirectly into water bodies without adequate treatment to remove harmful compounds.

	Water pollution affects plants and organisms living in these bodies of water.
Radiation	In physics, radiation is a process in which energetic particles or energetic waves travel through a medium or space. Two types of radiation are commonly differentiated in the way they interact with normal chemical matter: ionizing and non-ionizing radiation. The word radiation is often colloquially used in reference to ionizing radiation but the term radiation may correctly also refer to non-ionizing radiation.
Radioactive waste	Radioactive wastes are wastes that contain radioactive material. Radioactive wastes are usually by-products of nuclear power generation and other applications of nuclear fission or nuclear technology, such as research and medicine. Radioactive waste is hazardous to most forms of life and the environment, and is regulated by government agencies in order to protect human health and the environment.
Nuclear fusion	Nuclear fusion is the process by which two or more atomic nuclei join together, or 'fuse', to form a single heavier nucleus. This is usually accompanied by the release or absorption of large quantities of energy. Fusion is the process that powers active stars, the hydrogen bomb and some experimental devices examining fusion power for electrical generation.
Sustainability	Sustainability is the capacity to endure. For humans, sustainability is the long-term maintenance of responsibility, which has environmental, economic, and social dimensions, and encompasses the concept of stewardship, the responsible management of resource use. In ecology, sustainability describes how biological systems remain diverse and productive over time, a necessary precondition for human well-being.

1. _____ is the removal of a forest or stand of trees where the land is thereafter converted to a nonforest use. Examples of _____ include conversion of forestland to farms, ranches, or urban use.

   The term _____ is often misused to describe any activity where all trees in an area are removed.

   a. Desertification
   b. Genetic pollution
   c. Deforestation
   d. Global distillation

2. _____ in general describes the extent to which time or effort is well used for the intended task or purpose. It is often used with the specific purpose of relaying the capability of a specific application of effort to produce a specific outcome effectively with a minimum amount or quantity of waste, expense, or unnecessary effort. '_____' has widely varying meanings in different disciplines.

   a. Egain Forecasting
   b. Efficiency
   c. Equivalence of direct radiation
   d. Insulated glazing

3. _____s are the preserved remains or traces of animals (also known as zoolites), plants, and other organisms from the remote past. The totality of _____s, both discovered and undiscovered, and their placement in fossiliferous (_____-containing) rock formations and sedimentary layers (strata) is known as the _____ record.

   The study of _____s across geological time, how they were formed, and the evolutionary relationships between taxa (phylogeny) are some of the most important functions of the science of paleontology.

   a. Fossil collecting
   b. Francevillian Group Fossil
   c. Geologic record
   d. Fossil

---

4. _____ is the process by which two or more atomic nuclei join together, or 'fuse', to form a single heavier nucleus. This is usually accompanied by the release or absorption of large quantities of energy. Fusion is the process that powers active stars, the hydrogen bomb and some experimental devices examining fusion power for electrical generation.

    a. Nuclear transmutation
    b. Photofission
    c. Post Irradiation Examination
    d. Nuclear fusion

5. _____ is the process by which an atomic nucleus of an unstable atom loses energy by emitting ionizing particles (ionizing radiation) A decay, or loss of energy, results when an atom with one type of nucleus, called the parent radionuclide, transforms to an atom with a nucleus in a different state, or to a different nucleus containing different numbers of nucleons.

    a. Juglone
    b. Peak water
    c. Pigou Club
    d. Radioactive decay

---

1. c
2. b
3. d
4. d
5. d

You can take the complete Chapter Practice Test

for Chapter 12. Nonrenewable Energy: Coal, Oil, Natural Gas, and Nuclear Fuels
on all key terms, persons, places, and concepts.

Online 99 Cents

http://www.epub89.16.20190.12.cram101.com/

Use www.Cram101.com for all your study needs

including Cram101's online interactive problem solving labs in chemistry, statistics, mathematics, and more.

Energy conservation

Efficiency

Sustainable design

Thermodynamics

Inertia

Carbon cycle

Carbon dioxide

Chaparral

Fossil

Global change

Air pollution

Pollution

Atmosphere

Ethanol

Water pollution

Hydroelectricity

Kinetic energy

Three Gorges Dam

Glen Canyon Dam

Fish ladder

Siltation

Solar energy

Active solar

Solar cell

Geothermal energy

Radioactive decay

Wind energy

Electrolysis

Fuel cell

Hydrogen fuel

Sustainability

Alternative energy

Building

Corn ethanol

Fly ash

Ecology

Restoration ecology

Algal bloom

_____ | Estuary _____

_____ | Invertebrate _____

_____ | Wastewater _____

_____ | Biochemical oxygen demand _____

_____ | Cultural eutrophication _____

_____ | Dead zone _____

_____ | Eutrophication _____

_____ | Nutrient _____

_____ | Indicator species _____

_____ | Sewage _____

_____ | Sewage treatment _____

_____ | Sludge _____

_____ | Feedlot _____

_____ | Lagoon _____

_____ | Arsenic _____

_____ | Carcinogen _____

_____ | Heavy metal _____

_____ | Green Revolution _____

_____ | Hormone _____

Chapter 13. Renewable Energy: Innovative Uses of Earth, Sun, Wind, and Water

_____ | Biodiversity

_____ | Biodiversity hotspot

_____ | Endocrine disruptor

_____ | Hotspot

_____ | Cuyahoga River

_____ | Polychlorinated biphenyl

_____ | Oil spill

_____ | Sediment

_____ | Nuclear power

_____ | Thermal pollution

_____ | Thermal shock

_____ | Clean Water Act

_____ | Noise pollution

CHAPTER HIGHLIGHTS: KEY TERMS, PEOPLE, PLACES, CONCEPTS
Chapter 13. Renewable Energy: Innovative Uses of Earth, Sun, Wind, and Water

193

Energy conservation	Energy conservation refers to efforts made to reduce energy consumption. Energy conservation can be achieved through increased efficient energy use, in conjunction with dreased energy consumption and/or reduced consumption from conventional energy sources. An energy conservation act was passed in 2001.
Efficiency	Efficiency in general describes the extent to which time or effort is well used for the intended task or purpose. It is often used with the specific purpose of relaying the capability of a specific application of effort to produce a specific outcome effectively with a minimum amount or quantity of waste, expense, or unnecessary effort. 'Efficiency' has widely varying meanings in different disciplines.
Sustainable design	Sustainable design. is the philosophy of designing physical objects, the built environment, and services to comply with the principles of economic, social, and ecological sustainability.
	Intentions
	The intention of sustainable design is to 'eliminate negative environmental impact completely through skillful, sensitive design'.
Thermodynamics	ImgProperty database
	img
	Thermodynamics is the branch of physical science concerned with heat and its relation to other forms of energy and work. It defines macroscopic variables (such as temperature, entropy, and pressure) that describe average properties of material bodies and radiation, and explains how they are related and by what laws they change with time. Thermodynamics does not describe the microscopic constituents of matter, and its laws can be derived from statistical mechanics.
Inertia	Inertia is the resistance of any physical object to a change in its state of motion or rest, or the tendency of an object to resist any change in its motion. The principle of inertia is one of the fundamental principles of classical physics which are used to describe the motion of matter and how it is affected by applied forces. Inertia comes from the Latin word, iners, meaning idle, or lazy.

Carbon cycle	The carbon cycle is the biogeochemical cycle by which carbon is exchanged among the biosphere, pedosphere, geosphere, hydrosphere, and atmosphere of the Earth. It is one of the most important cycles of the earth and allows for carbon to be recycled and reused throughout the biosphere and all of its organisms.
	The carbon cycle was initially discovered by Joseph Priestley and Antoine Lavoisier, and popularized by Humphry Davy.
Carbon dioxide	Carbon dioxide is a naturally occurring chemical compound composed of two oxygen atoms covalently bonded to a single carbon atom. It is a gas at standard temperature and pressure and exists in Earth's atmosphere in this state, as a trace gas at a concentration of 0.039% by volume.
	As part of the carbon cycle known as photosynthesis, plants, algae, and cyanobacteria absorb carbon dioxide, light, and water to produce carbohydrate energy for themselves and oxygen as a waste product.
Chaparral	Chaparral is a shrubland or heathland plant community found primarily in the U.S. state of California and in the northern portion of the Baja California peninsula, Mexico. It is shaped by a Mediterranean climate (mild, wet winters and hot dry summers) and wildfire, having summer drought-tolerant plants with hard sclerophyllous evergreen leaves, as contrasted with the associated soft leaved, drought deciduous, scrub community of Coastal sage scrub, found below the chaparral biome. Chaparral covers 5% of the state of California, and associated Mediterranean scrubland an additional 3.5%.
Fossil	Fossils are the preserved remains or traces of animals (also known as zoolites), plants, and other organisms from the remote past. The totality of fossils, both discovered and undiscovered, and their placement in fossiliferous (fossil-containing) rock formations and sedimentary layers (strata) is known as the fossil record.

	The study of fossils across geological time, how they were formed, and the evolutionary relationships between taxa (phylogeny) are some of the most important functions of the science of paleontology.
Global change	Global change refers to planetary-scale changes in the Earth system. The system consists of the land, oceans, atmosphere, poles, life, the planet's natural cycles and deep Earth processes. These constituent parts influence one another.
Air pollution	Air pollution is the introduction of chemicals, particulate matter, or biological materials that cause harm or discomfort to humans or other living organisms, or cause damage to the natural environment or built environment, into the atmosphere. The atmosphere is a complex dynamic natural gaseous system that is essential to support life on planet Earth. Stratospheric ozone depletion due to air pollution has long been recognized as a threat to human health as well as to the Earth's ecosystems.
Pollution	Pollution is the introduction of contaminants into a natural environment that causes instability, disorder, harm or discomfort to the ecosystem i.e. physical systems or living organisms. Pollution can take the form of chemical substances or energy, such as noise, heat or light. Pollutants, the components of pollution, can be either foreign substances/energies or naturally occurring contaminants.
Atmosphere	The standard atmosphere (symbol: atm) is an international reference pressure defined as 101325 Pa and formerly used as unit of pressure. For practical purposes it has been replaced by the bar which is 10^5 Pa. The difference of about 1% is not significant for many applications, and is within the error range of common pressure gges.
Ethanol	Ethanol, pure alcohol, grain alcohol, or drinking alcohol, is a volatile, flammable, colorless liquid. It is a psychoactive drug and one of the oldest recreational drugs. Best known as the type of alcohol found in alcoholic beverages, it is also used in thermometers, as a solvent, and as a fuel.

Water pollution	Water pollution is the contamination of water bodies (e.g. lakes, rivers, oceans, aquifers and groundwater). Water pollution occurs when pollutants are discharged directly or indirectly into water bodies without adequate treatment to remove harmful compounds. Water pollution affects plants and organisms living in these bodies of water.
Hydroelectricity	Hydroelectricity is the term referring to electricity generated by hydropower; the production of electrical power through the use of the gravitational force of falling or flowing water. It is the most widely used form of renewable energy. Once a hydroelectric complex is constructed, the project produces no direct waste, and has a considerably lower output level of the greenhouse gas carbon dioxide (CO_2) than fossil fuel powered energy plants.
Kinetic energy	The kinetic energy of an object is the energy which it possesses due to its motion. It is defined as the work needed to accelerate a body of a given mass from rest to its stated velocity. Having gained this energy during its acceleration, the body maintains this kinetic energy unless its speed changes.
Three Gorges Dam	The Three Gorges Dam is a hydroelectric dam that spans the Yangtze River by the town of Sandouping, located in the Yiling District of Yichang, in Hubei province, China. The Three Gorges Dam is the world's largest power station in terms of installed capacity (21,000 MW) but is second to Itaipu Dam with regard to the generation of electricity annually. The dam body was completed in 2006. Except for a ship lift, the originally planned components of the project were completed on October 30, 2008, when the 26th turbine in the shore plant began commercial operation.
Glen Canyon Dam	Glen Canyon Dam is a concrete arch dam on the Colorado River in northern Arizona in the United States, just north of Page. The dam was built to provide hydroelectricity and flow regulation from the upper Colorado River Basin to the lower. Its reservoir is called Lake Powell, and is the second largest artificial lake in the country, extending upriver well into Utah.

Fish ladder	A fish ladder, fish pass or fish steps, is a structure on or around artificial barriers (such as dams and locks) to facilitate diadromous fishes' natural migration. Most fishways enable fish to pass around the barriers by swimming and leaping up a series of relatively low steps (hence the term ladder) into the waters on the other side. The velocity of water falling over the steps has to be great enough to attract the fish to the ladder, but it cannot be so great that it washes fish back downstream or exhausts them to the point of inability to continue their journey upriver.
Siltation	Siltation is the pollution of water by fine particulate terrestrial clastic material, with a particle size dominated by silt or clay. It refers both to the increased concentration of suspended sediments, and to the increased accumulation (temporary or permanent) of fine sediments on bottoms where they are undesirable. Siltation is most often caused by soil erosion or sediment spill.
Solar energy	Solar energy, radiant light and heat from the sun, has been harnessed by humans since ancient times using a range of ever-evolving technologies. Solar radiation, along with secondary solar-powered resources such as wind and wave power, hydroelectricity and biomass, account for most of the available renewable energy on earth. Only a minuscule fraction of the available solar energy is used.
Active solar	Active solar technologies are employed to convert solar energy into usable light, heat, cause air-movement for ventilation or cooling, or store heat for future use. Active solar uses electrical or mechanical equipment, such as pumps and fans, to increase the usable heat in a system. Solar energy collection and utilization systems that do not use external energy, like a solar chimney, are classified as passive solar technologies.
Solar cell	A solar cell is a solid state electrical device that converts the energy of light directly into electricity by the photovoltaic effect. Assemblies of cells used to make solar modules which are used to capture energy from sunlight, are known as solar panels. The energy generated from these solar modules, referred to as solar power, is an example of solar energy.
Geothermal energy	Geothermal energy is thermal energy generated and stored in the Earth. Thermal energy is the energy that determines the temperature of matter. Earth's geothermal energy originates from the original formation of the planet (20%) and from radioactive decay of minerals (80%).

Radioactive decay	Radioactive decay is the process by which an atomic nucleus of an unstable atom loses energy by emitting ionizing particles (ionizing radiation) A decay, or loss of energy, results when an atom with one type of nucleus, called the parent radionuclide, transforms to an atom with a nucleus in a different state, or to a different nucleus containing different numbers of nucleons.
Wind energy	Wind energy is the kinetic energy of the air in motion. Total wind energy flowing through an imaginary area A during the time t is: $E = A \cdot v \cdot$
Electrolysis	In chemistry and manufacturing, electrolysis is a method of using a direct electric current (DC) to drive an otherwise non-spontaneous chemical reaction. Electrolysis is commercially highly important as a stage in the separation of elements from naturally occurring sources such as ores using an electrolytic cell. History The word electrolysis comes from the Greek ?λεκτρον [lýsis] 'dissolution'.
Fuel cell	A fuel cell is a device that converts the chemical energy from a fuel into electricity through a chemical reaction with oxygen or another oxidizing agent. Hydrogen is the most common fuel, but hydrocarbons such as natural gas and alcohols like methanol are sometimes used. Fuel cells are different from batteries in that they require a constant source of fuel and oxygen to run, but they can produce electricity continually for as long as these inputs are supplied.
Hydrogen fuel	Hydrogen fuel is an eco-friendly fuel which uses electrochemical cells, or combustion in internal engines, to power vehicles and electric devices. It is also used in the propulsion of spacecraft and can potentially be mass produced and commercialized for passenger vehicles and aircraft. Chemistry Hydrogen is the first element on the periodic table, making it the lightest element on earth.

Sustainability	Sustainability is the capacity to endure. For humans, sustainability is the long-term maintenance of responsibility, which has environmental, economic, and social dimensions, and encompasses the concept of stewardship, the responsible management of resource use. In ecology, sustainability describes how biological systems remain diverse and productive over time, a necessary precondition for human well-being.
Alternative energy	Alternative energy is an umbrella term that refers to any source of usable energy intended to replace fuel sources without the undesired consequences of the replaced fuels. The term 'alternative' presupposes a set of undesirable energy technologies against which 'alternative energies' are contrasted. As such, the list of energy technologies excluded is an indicator of which problems the alternative technologies are intended to address.
Building	In mathematics, a building (also Tits building, Bruhat-Tits building, finite projective planes, and Riemannian symmetric spaces. Initially introduced by Jacques Tits as a means to understand the structure of exceptional groups of Lie type, the theory has also been used to study the geometry and topology of homogeneous spaces of p-adic Lie groups and their discrete subgroups of symmetries, in the same way that trees have been used to study free groups. Overview The notion of a building was invented by Jacques Tits as a means of describing simple algebraic groups over an arbitrary field.
Corn ethanol	Corn ethanol is ethanol produced from corn as a biomass through industrial fermentation, chemical processing and distillation. Corn is the main feedstock used for producing ethanol fuel in the United States and it is mainly used as an oxygenate to gasoline in the form of low-level blends, and to a lesser extent, as fuel for E85 flex-fuel vehicles.
Fly ash	Fly ash is one of the residues generated in combustion, and comprises the fine particles that rise with the flue gases. Ash which does not rise is termed bottom ash. In an industrial context, fly ash usually refers to ash produced during combustion of coal.

Ecology	Ecology is the scientific study of the relations that living organisms have with respect to each other and their natural environment. Variables of interest to ecologists include the composition, distribution, amount (biomass), number, and changing states of organisms within and among ecosystems. Ecosystems are hierarchical systems that are organized into a graded series of regularly interacting and semi-independent parts (e.g., species) that aggregate into higher orders of complex integrated wholes (e.g., communities).
Restoration ecology	Restoration ecology is the scientific study and practice of renewing and restoring degraded, damaged, or destroyed ecosystems and habitats in the environment by active human intervention and action. Restoration ecology emerged as a separate field in ecology in the 1980s.

History

Land managers, laypeople, and stewards have been practicing restoration for many hundreds, if not thousands of years, yet the scientific field of 'restoration ecology' was first identified and coined in the late 1980s by John Aber and William Jordan. |
| Algal bloom | An algal bloom is a rapid increase or accumulation in the population of algae in an aquatic system. Algal blooms may occur in freshwater as well as marine environments. Typically, only one or a small number of phytoplankton species are involved, and some blooms may be recognized by discoloration of the water resulting from the high density of pigmented cells. |
| Estuary | An estuary is a partly enclosed coastal body of water with one or more rivers or streams flowing into it, and with a free connection to the open sea.

Estuaries form a transition zone between river environments and ocean environments and are subject to both marine influences, such as tides, waves, and the influx of saline water; and riverine influences, such as flows of fresh water and sediment. The inflow of both seawater and freshwater provide high levels of nutrients in both the water column and sediment, making estuaries among the most productive natural habitats in the world. |

Invertebrate	An invertebrate is an animal without a backbone. The group includes 97% of all animal species - all animals except those in the chordate subphylum Vertebrata (fish, amphibians, reptiles, birds, and mammals). Invertebrates form a paraphyletic group.
Wastewater	Waste Water is any water that has been adversely affected in quality by anthropogenic influence. It comprises liquid waste discharged by domestic residences, commercial properties, industry, and/or agriculture and can encompass a wide range of potential contaminants and concentrations. In the most common usage, it refers to the municipal wastewater that contains a broad spectrum of contaminants resulting from the mixing of wastewaters from different sources.
Biochemical oxygen demand	Biochemical oxygen demand or B.O.D. is the amount of dissolved oxygen needed by aerobic biological organisms in a y of water to break down organic material present in a given water sample at certain temperature over a specific time period. The term also refers to a chemical procedure for determining this amount. This is not a precise quantitative test, although it is widely used as an indication of the organic quality of water.
Cultural eutrophication	Cultural eutrophication is the pross that speeds up natural eutrophication because of human activity. Due to clearing of land and building of towns and cities, land runoff is aclerated and more nutrients such as phosphates and nitrate are supplied to lakes and rivers, and then to coastal estuaries and bays. Extra nutrients are also supplied by treatment plants, golf courses, fertilizers, and farms.
Dead zone	Dead zones are hypoxic (low-oxygen) areas in the world's oceans, the observed incidences of which have been increasing since oceanographers began noting them in the 1970s. These occur near inhabited coastlines, where aquatic life is most concentrated. (The vast middle portions of the oceans which naturally have little life are not considered 'dead zones'). The term can also be applied to the identical phenomenon in large lakes.
Eutrophication	Eutrophication, is the ecosystem response to the addition of artificial or natural substances, such as nitrates and phosphates, through fertilizers or sewage, to an aquatic system. One example is the 'bloom' or great increase of phytoplankton in a water body as a response to increased levels of nutrients. Negative environmental effects include hypoxia, the depletion of oxygen in the water, which induces reductions in specific fish and other animal populations.

Nutrient	A nutrient is a chemical that an organism needs to live and grow or a substance used in an organism's metabolism which must be taken in from its environment. They are used to build and repair tissues, regulate body processes and are converted to and used as energy. Methods for nutrient intake vary, with animals and protists consuming foods that are digested by an internal digestive system, but most plants ingest nutrients directly from the soil through their roots or from the atmosphere.
Indicator species	An indicator species is any biological species that defines a trait or characteristic of the environment. For example, a species may delineate an ecoregion or indicate an environmental condition such as a disease outbreak, pollution, species competition or climate change. Indicator species can be among the most sensitive species in a region, and sometimes act as an early warning to monitoring biologists.
Sewage	Sewage is water-carried waste, in solution or suspension, that is intended to be removed from a community. Also known as wastewater, it is more than 99% water and is characterized by volume or rate of flow, physical condition, chemical constituents and the bacteriological organisms that it contains. In loose American English usage, the terms 'sewage' and 'sewerage' are sometimes interchanged.
Sewage treatment	Sewage treatment, is the process of removing contaminants from waewater and household sewage, both runoff (effluents) and domeic. It includes physical, chemical, and biological processes to remove physical, chemical and biological contaminants. Its objective is to produce an environmentally-safe fluid wae ream (or treated effluent) and a solid wae (or treated sludge) suitable for disposal or reuse (usually as farm fertilizer).
Sludge	Sludge refers to the residual, semi-solid material left from industrial wastewater, or sewage treatment processes. It can also refer to the settled suspension obtained from conventional drinking water treatment, and numerous other industrial processes. The term is also sometimes used as a generic term for solids separated from suspension in a liquid; this 'soupy' material usually contains significant quantities of 'interstitial' water (between the solid particles).
Feedlot	A feedlot is a type of animal feeding operation (AFO) which is used in factory farming for finishing livestock, notably beef cattle, but also swine, horses, sheep, turkeys, chickens or ducks, prior to slaughter. Large beef feedlots are called Concentrated Animal Feeding Operations (CAFOs). They may contain thousands of animals in an array of pens.

Lagoon	A lagoon is a body of shallow sea water or brackish water separated from the sea by some form of barrier. The EU's habitat directive defines lagoons as 'expanses of shallow coastal salt water, of varying salinity or water volume, wholly or partially separated from the sea by sand banks or shingle, or, less frequently, by rocks. Salinity may vary from brackish water to hypersalinity depending on rainfall, evaporation and through the addition of fresh seawater from storms, temporary flooding by the sea in winter or tidal exchange'.
Arsenic	Arsenic is a chemical element with the symbol As, atomic number 33 and relative atomic mass 74.92. Arsenic occurs in many minerals, usually in conjunction with sulfur and metals, and also as a pure elemental crystal. It was first documented by Albertus Magnus in 1250. Arsenic is a metalloid. It can exist in various allotropes, although only the grey form has important use in industry.
Carcinogen	A carcinogen is any substance, radionuclide or radiation, that is an agent directly involved in causing cancer. This may be due to the ability to damage the genome or to the disruption of cellular metabolic processes. Several radioactive substances are considered carcinogens, but their carcinogenic activity is attributed to the radiation, for example gamma rays and alpha particles, which they emit.
Heavy metal	A heavy metal is a member of a loosely-defined subset of elements that exhibit metallic properties. It mainly includes the transition metals, some metalloids, lanthanides, and actinides. Many different definitions have been proposed--some based on density, some on atomic number or atomic weight, and some on chemical properties or toxicity.
Green Revolution	Green Revolution refers to a series of research, development, and technology transfer initiatives, occurring between the 1940s and the late 1970s, that increased agriculture production around the world, beginning most markedly in the late 1960s. The initiatives, led by Norman Borlaug, the 'Father of the Green Revolution' credited with saving over a billion people from starvation, involved the development of high-yielding varieties of cereal grains, expansion of irrigation infrastructure, modernization of management techniques, distribution of hybridized seeds, synthetic fertilizers, and pesticides to farmers.

The term 'Green Revolution' was first used in 1968 by former United States Agency for International Development (USAID) director William Gaud, who noted the spread of the new technologies and said,

These and other developments in the field of agriculture contain the makings of a new revolution.

Hormone	A hormone is a chemical released by a cell or a gland in one part of the body that sends out messages that affect cells in other parts of the organism. Only a small amount of hormone is required to alter cell metabolism. In essence, it is a chemical messenger that transports a signal from one cell to another.
Biodiversity	Biodiversity is the degree of variation of life forms within a given species, ecosystem, biome, or an entire planet. Biodiversity is a measure of the health of ecosystems. Biodiversity is in part a function of climate.
Biodiversity hotspot	A biodiversity hotspot is a biogeographic region with a significant reservoir of biodiversity that is under threat from humans.

The concept of biodiversity hotspots was originated by Norman Myers in two articles in 'The Environmentalist' (1988 ' 1990), revised after thorough analysis by Myers and others in 'Hotspots: Earth's Biologically Richest and Most Endangered Terrestrial Ecoregions'.

To qualify as a biodiversity hotspot on Myers 2000 edition of the hotspot-map, a region must meet two strict criteria: it must contain at least 0.5% or 1,500 species of vascular plants as endemics, and it has to have lost at least 70% of its primary vegetation.

Endocrine disruptor	Endocrine disruptors are chemicals that interfere with endocrine (or hormone system) in animals, including humans. These disruptions can cause cancerous tumors, birth defects, and other developmental disorders. Specifically, they are known to cause learning disabilities, severe attention deficit disorder, cognitive and brain development problems, deformations of the body (including limbs); sexual development problems, feminizing of males or masculine effects on females, etc.

Hotspot	The places known as hotspots or hot spots in geology are volcanic regions thought to be fed by underlying mantle that is anomalously hot compared with the mantle elsewhere. They may be on, near to, or far from tectonic plate boundaries. There are two hypotheses to explain them.
Cuyahoga River	The Cuyahoga River is located in Northeast Ohio in the United States. Outside of Ohio, the river is most famous for being 'the river that caught fire', helping to spur the environmental movement in the late 1960s. Native Americans called this winding water 'Cuyahoga,' which means 'ooked river' in the Iroquois language.
Polychlorinated biphenyl	A polychlorinated biphenyl is any of the 209 configurations of organochlorides with 2 to 10 chlorine atoms attached to biphenyl, which is a molecule composed of two benzene rings. The chemical formula for a PCB is $C_{12}H_{10-x}Cl_x$. 130 of the 209 different PCB arrangements and orientations are used commercially.
Oil spill	An oil spill is the release of a liquid petroleum hydrocarbon into the environment, especially marine areas, due to human activity, and is a form of pollution. The term is mtly used to describe marine oil spills, where oil is released into the ocean or coastal waters. Oil spills may be due to releases of crude oil from tankers, offshore platforms, drilling rigs and wells, as well as spills of refined petroleum products (such as gasoline, diesel) and their by-products, heavier fuels used by large ships such as bunker fuel, or the spill of any oily refuse or waste oil.
Sediment	Sediment is naturally-occurring material that is broken down by processes of weathering and erosion, and is subsequently transported by the action of fluids such as wind, water, or ice, and/or by the force of gravity acting on the particle itself. Sediments are most often transported by water (fluvial processes) transported by wind (aeolian processes) and glaciers. Beach sands and river channel deposits are examples of fluvial transport and deposition, though sediment also often settles out of slow-moving or standing water in lakes and oceans.
Nuclear power	Nuclear power is the use of sustained nuclear fission to generate heat and electricity. Nuclear power plants provide about 6% of the world's energy and 13-14% of the world's electricity, with the U.S., France, and Japan together accounting for about 50% of nuclear generated electricity. In 2007, the IAEA reported there were 439 nuclear power reactors in operation in the world, operating in 31 countries.

Thermal pollution	Thermal pollution is the degradation of water quality by any process that changes ambient water temperature.
	A common cause of thermal pollution is the use of water as a coolant by power plants and industrial manufacturers. When water used as a coolant is returned to the natural environment at a higher temperature, the change in temperature decreases oxygen supply, and affects ecosystem composition.
Thermal shock	Thermal shock is the name given to extreme temperature difference (gradient) across an object, which can result in cracking and/or breaking. Glass and ceramic objec are particularly vulnerable to this form of failure, due to their low toughness and low thermal conductivity. However, they are used in many high temperature applications due to their high melting point.
Clean Water Act	The Clean Water Act is the primary federal law in the United States governing water pollution. Commonly abbreviated as the Clean Water Act, the act established the goals of eliminating releases of high amounts of toxic substances into water, eliminating additional water pollution by 1985, and ensuring that surface waters would meet standards necessary for human sports and recreation by 1983.
	The principal body of law currently in effect is based on the Federal Water Pollution Control Amendments of 1972 and was significantly expanded from the Federal Water Pollution Control Amendments of 1948. Major amendments were enacted in the Clean Water Act of 1977 and the Water Quality Act of 1987.
Noise pollution	Noise pollution is excessive, displeasing human, animal, or machine-created environmental noise that disrupts the activity or balance of human or animal life. The word noise may be from the Latin word nauseas, metaphorically meaning disgust or discomfort.
	The source of most outdoor noise worldwide is mainly construction and transportation systems, including motor vehicle noise, aircraft noise, and rail noise.

---

1. _____, is the ecosystem response to the addition of artificial or natural substances, such as nitrates and phosphates, through fertilizers or sewage, to an aquatic system. One example is the 'bloom' or great increase of phytoplankton in a water body as a response to increased levels of nutrients. Negative environmental effects include hypoxia, the depletion of oxygen in the water, which induces reductions in specific fish and other animal populations.

   a. Extended aeration
   b. In situ chemical reduction
   c. Indicator bacteria
   d. Eutrophication

2. _____ is the capacity to endure. For humans, _____ is the long-term maintenance of responsibility, which has environmental, economic, and social dimensions, and encompasses the concept of stewardship, the responsible management of resource use. In ecology, _____ describes how biological systems remain diverse and productive over time, a necessary precondition for human well-being.

   a. Back-to-the-land movement
   b. Sustainability
   c. Best of the Web Directory
   d. Biocapacity

3. _____. is the philosophy of designing physical objects, the built environment, and services to comply with the principles of economic, social, and ecological sustainability.

   Intentions
   The intention of _____ is to 'eliminate negative environmental impact completely through skillful, sensitive design'.

   a. Sustainable design
   b. Sustainable event management
   c. Sustainable forest management
   d. Sustainable furniture design

4. A _____ is any substance, radionuclide or radiation, that is an agent directly involved in causing cancer. This may be due to the ability to damage the genome or to the disruption of cellular metabolic processes. Several radioactive substances are considered _____s, but their carcinogenic activity is attributed to the radiation, for example gamma rays and alpha particles, which they emit.

---

a. Juglone
b. Carcinogen
c. Iprodione
d. Wave-cut platform

5. _____ refers to efforts made to reduce energy consumption. _____ can be achieved through increased efficient energy use, in conjunction with dreased energy consumption and/or reduced consumption from conventional energy sources. An _____ act was passed in 2001.

a. Ethanol fuel energy balance
b. Energy conservation
c. Odyssey
d. Endangered Species Act

---

1. d
2. b
3. a
4. b
5. b

You can take the complete Chapter Practice Test

for Chapter 13. Renewable Energy: Innovative Uses of Earth, Sun, Wind, and Water
on all key terms, persons, places, and concepts.

Online 99 Cents

http://www.epub89.16.20190.13.cram101.com/

Use www.Cram101.com for all your study needs

including Cram101's online interactive problem solving labs in chemistry, statistics, mathematics, and more.

CHAPTER OUTLINE: KEY TERMS, PEOPLE, PLACES, CONCEPTS
Chapter 14
Air Pollution: Causes, Effects, and Stratospheric Ozone Depletion

211

Air pollution

Pollution

Atmosphere

Carbon dioxide

Carbon monoxide

Global change

Sulfur

Bioaccumulation

Primary

Pollutant

Great Smoky Mountains

Ozone

Ozone layer

Lignite

Nitric acid

Sulfuric acid

Nuclear power

Baghouse

Electrostatics

_____ | Electrostatic precipitator

_____ | Scrubber

_____ | Acid rain

_____ | Chlorofluorocarbon

_____ | Radiation

_____ | Catalyst

_____ | Montreal Protocol

_____ | Water pollution

_____ | Asbestos

_____ | Radon

_____ | Carcinogen

_____ | Biogeography

_____ | Sustainability

CHAPTER HIGHLIGHTS: KEY TERMS, PEOPLE, PLACES, CONCEPTS
Chapter 14. Air Pollution: Causes, Effects, and Stratospheric Ozone Depletion

213

Air pollution	Air pollution is the introduction of chemicals, particulate matter, or biological materials that cause harm or discomfort to humans or other living organisms, or cause damage to the natural environment or built environment, into the atmosphere.
	The atmosphere is a complex dynamic natural gaseous system that is essential to support life on planet Earth. Stratospheric ozone depletion due to air pollution has long been recognized as a threat to human health as well as to the Earth's ecosystems.
Pollution	Pollution is the introduction of contaminants into a natural environment that causes instability, disorder, harm or discomfort to the ecosystem i.e. physical systems or living organisms. Pollution can take the form of chemical substances or energy, such as noise, heat or light. Pollutants, the components of pollution, can be either foreign substances/energies or naturally occurring contaminants.
Atmosphere	The standard atmosphere (symbol: atm) is an international reference pressure defined as 101325 Pa and formerly used as unit of pressure. For practical purposes it has been replaced by the bar which is 10^5 Pa. The difference of about 1% is not significant for many applications, and is within the error range of common pressure gges.
Carbon dioxide	Carbon dioxide is a naturally occurring chemical compound composed of two oxygen atoms covalently bonded to a single carbon atom. It is a gas at standard temperature and pressure and exists in Earth's atmosphere in this state, as a trace gas at a concentration of 0.039% by volume.
	As part of the carbon cycle known as photosynthesis, plants, algae, and cyanobacteria absorb carbon dioxide, light, and water to produce carbohydrate energy for themselves and oxygen as a waste product.
Carbon monoxide	Carbon monoxide also called carbonous oxide, is a colorless, odorless and tasteless gas which is slightly lighter than air. It is highly toxic to humans and animals in higher quantities, although it is also produced in normal animal metabolism in low quantities, and is thought to have some normal biological functions.

	It consists of one carbon atom and one oxygen atom, connected by a triple bond which consists of two covalent bonds as well as one dative covalent bond.
Global change	Global change refers to planetary-scale changes in the Earth system. The system consists of the land, oceans, atmosphere, poles, life, the planet's natural cycles and deep Earth processes. These constituent parts influence one another.
Sulfur	Sulfur or sulphur is the chemical element with atomic number 16. In the periodic table it is represented by the symbol S. It is an abundant, multivalent non-metal. Under normal conditions, sulfur atoms form cyclic octatomic molecules with chemical formula S_8. Elemental sulfur is a bright yellow crystalline solid when at room temperature.
Bioaccumulation	Bioaccumulation refers to the accumulation of substances, such as pesticides, or other organic chemicals in an organism. Bioaccumulation occurs when an organism absorbs a toxic substance at a rate greater than that at which the substance is lost. Thus, the longer the biological half-life of the substance the greater the risk of chronic poisoning, even if environmental levels of the toxin are not very high.
Primary	A primary (or gravitational primary) is the main physical body of a gravitationally bound, multi-object system. This body contributes most of the mass of that system and will generally be located near its center of mass. In the solar system, the Sun is the primary for all objects that orbit around it.
Pollutant	A pollutant is a waste material that pollutes air, water or soil, and is the cause of pollution. Three factors determine the severity of a pollutant: its chemical nature, its concentration and its persistence. Some pollutants are biodegradable and therefore will not persist in the environment in the long term.

Great Smoky Mountains	The Great Smoky Mountains are a mountain range rising along the Tennessee-North Carolina border in the southeastern United States. They are a subrange of the Appalachian Mountains, and form part of the Blue Ridge Physiographic Province. The range is sometimes called the Smoky or Smokey Mountains, and the name is commonly shortened to the Smokies.
Ozone	Ozone or trioxygen, is a triatomic molecule, consisting of three oxygen atoms. It is an allotrope of oxygen that is much less stable than the diatomic allotrope (O_2). Ozone in the lower atmosphere is an air pollutant with harmful effects on the respiratory systems of animals and will burn sensitive plants; however, the ozone layer in the upper atmosphere is beneficial, preventing damaging ultraviolet light from reaching the Earth's surface.
Ozone layer	The ozone layer is a layer in Earth's atmosphere which contains relatively high concentrations of ozone (O_3). This layer absorbs 97-99% of the Sun's high frequency ultraviet light, which potentially damages the life forms on Earth. It is mainly located in the lower portion of the stratosphere from approximately 20 to 30 kilometres (12 to 19 mi) above Earth, though the thickness varies seasonally and geographically.
Lignite	Lignite, often referred to as brown coal, or Rosebud coal by Northern Pacific Railroad, is a soft brown fuel with characteristics that put it somewhere between coal and peat. It is considered the lowest rank of coal; it is mined in Greece, Germany, Poland, Serbia, Russia, the United States, India, Australia and many other parts of Europe and it is used almost exclusively as a fuel for steam-electric power generation. Up to 50% of Greece's electricity and 24.6% of Germany's comes from lignite power plants.
Nitric acid	Nitric acid also known as aqua fortis and spirit of nitre, is a highly corrosive and toxic strong mineral acid which is normally colorless but tends to acquire a yellow cast due to the accumulation of oxides of nitrogen if long-stored. Ordiry nitric acid has a concentration of 68%. When the solution contains more than 86% HNO_3, it is referred to as fuming nitric acid.
Sulfuric acid	Sulfuric acid is a highly corrosive strong mineral acid with the molecular formula H_2SO_4. The historical name of this acid is oil of vitriol. It is a colorless to slightly yellow viscous liquid and is soluble in water at all concentrations.

Nuclear power	Nuclear power is the use of sustained nuclear fission to generate heat and electricity. Nuclear power plants provide about 6% of the world's energy and 13-14% of the world's electricity, with the U.S., France, and Japan together accounting for about 50% of nuclear generated electricity. In 2007, the IAEA reported there were 439 nuclear power reactors in operation in the world, operating in 31 countries.
Baghouse	A baghouse or fabric filter (FF) is an air pollution control device that removes particulates out of air or gas released from commercial processes or combustion for electricity generation. Power plants, steel mills, pharmaceutical producers, food manufactures, chemical producers and other industrial companies often use baghouses to control emission of air pollutants. Baghouses came into widespread use in the late 1970s after the invention of high-temperature fabrics (for use in the filter media) capable of withstanding temperatures over 350°F.
	Unlike electrostatic precipitators, where performance can vary significantly depending on process and electrical conditions, functioning baghouses typically have a particulate collection efficiency of 99% or better, even when particle size is very small.
Electrostatics	Electrostatics is the branch of physics that deals with the phenomena and properties of stationary or slow-moving (without acceleration) electric charges.
	Since classical antiquity, it was known that some materials such as amber attract lightweight particles after rubbing. The Greek word for amber, ?λεκτρον electron, was the source of the word 'electricity'.
Electrostatic precipitator	An electrostatic precipitator or electrostatic air cleaner is a particulate collection device that removes particles from a flowing gas (such as air) using the force of an induced electrostatic charge. Electrostatic precipitators are highly efficient filtration devices that minimally impede the flow of gases through the device, and can easily remove fine particulate matter such as dust and smoke from the air stream. In contrast to wet scrubbers which apply energy directly to the flowing fluid medium, an ESP applies energy only to the particulate matter being collected and therefore is very efficient in its consumption of energy (in the form of electricity).

Scrubber	'Scrubber' systems are a diverse group of air pollution control devices that can be used to remove some particulates and/or gases from industrial exhaust streams. Traditionally, the term 'scrubber' has referred to pollution control devices that use liquid to wash unwanted pollutants from a gas stream. Recently, the term is also used to describe systems that inject a dry reagent or slurry into a dirty exhaust stream to 'wash out' acid gases.
Acid rain	Acid rain is a rain or any other form of precipitation that is unusually acidic, meaning that it possesses elevated levels of hydrogen ions (low pH). It can have hmful effects on plants, aquatic animals, and infrastructure. Acid rain is caused by emissions of cbon dioxide, sulfur dioxide and nitrogen oxides which react with the water molecules in the atmosphere to produce acids.
Chlorofluorocarbon	A chlorofluorocarbon is an organic compound that contains carbon, chlorine, and fluorine, produced as a volatile derivative of methane and ethane. A common subclass are the hydrochlorofluorocarbons (HCFCs), which contain hydrogen, as well. They are also commonly known by the DuPont trade name Freon.
Radiation	In physics, radiation is a process in which energetic particles or energetic waves travel through a medium or space. Two types of radiation are commonly differentiated in the way they interact with normal chemical matter: ionizing and non-ionizing radiation. The word radiation is often colloquially used in reference to ionizing radiation but the term radiation may correctly also refer to non-ionizing radiation.
Catalyst	Catalysis is the change in rate of a chemical reaction due to the participation of a substance called a catalyst. Unlike other reagents that participate in the chemical reaction, a catalyst is not consumed by the reaction itself. A catalyst may participate in multiple chemical transformations.
Montreal Protocol	The Montreal Protocol on Substances That Deplete the Ozone Layer (a protocol to the Vienna Convention for the Protection of the Ozone Layer) is an international treaty designed to protect the ozone layer by phasing out the production of numerous substances believed to be responsible for ozone depletion. The treaty was opened for signature on September 16, 1987, and entered into force on January 1, 1989, followed by a first meeting in Helsinki, May 1989. Since then, it has undergone seven revisions, in 1990 (London), 1991 (Nairobi), 1992 (Copenhagen), 1993 (Bangkok), 1995 (Vienna), 1997 (Montreal), and 1999 (Beijing). It is believed that if the international agreement is adhered to, the ozone layer is expected to recover by 2050. Due to its widespread adoption and implementation it has been hailed as an example of exceptional international co-operation, with Kofi Annan quoted as saying that 'perhaps the single most successful international agreement to date has been the Montreal Protocol'.

Water pollution	Water pollution is the contamination of water bodies (e.g. lakes, rivers, oceans, aquifers and groundwater). Water pollution occurs when pollutants are discharged directly or indirectly into water bodies without adequate treatment to remove harmful compounds.
	Water pollution affects plants and organisms living in these bodies of water.
Asbestos	Asbestos is a set of six naturally occurring silicate minerals exploited commercially for their desirable physical properties. They all have in common their asbestiform habit, long, (1:20) thin fibrous crystals. The inhalation of asbestos fibers can cause serious illnesses, including malignant lung cancer, mesothelioma (a formerly rare cancer strongly associated with exposure to amphibole asbestos), and asbestosis (a type of pneumoconiosis).
Radon	Radon is a chemical element with the atomic number 86, and is represented by the symbol Rn. It is a radioactive, colorless, odorless, tasteless noble gas, occurring naturally as the decay product of uranium or thorium. Its most stable isotope, ^{222}Rn, has a half-life of 3.8 days.
Carcinogen	A carcinogen is any substance, radionuclide or radiation, that is an agent directly involved in causing cancer. This may be due to the ability to damage the genome or to the disruption of cellular metabolic processes. Several radioactive substances are considered carcinogens, but their carcinogenic activity is attributed to the radiation, for example gamma rays and alpha particles, which they emit.
Biogeography	Biogeography is the study of the distribution of species (biology), organisms, and ecosystems in space and through geological time. Organisms and biological communities vary in a highly regular fashion along geographic gradients of latitude, elevation, isolation and habitat area.
	Knowledge of spatial variation in the numbers and types of organisms is as vital to us today as it was to our early human ancestors, as we adapt to heterogeneous but geographically predictable environments.

| Sustainability | Sustainability is the capacity to endure. For humans, sustainability is the long-term maintenance of responsibility, which has environmental, economic, and social dimensions, and encompasses the concept of stewardship, the responsible management of resource use. In ecology, sustainability describes how biological systems remain diverse and productive over time, a necessary precondition for human well-being. |

---

1. _____ is the contamination of water bodies (e.g. lakes, rivers, oceans, aquifers and groundwater). _____ occurs when pollutants are discharged directly or indirectly into water bodies without adequate treatment to remove harmful compounds.

   _____ affects plants and organisms living in these bodies of water.

   a. Water pollution
   b. Bioassay
   c. Bioretention
   d. Biosurvey

2. _____ is the capacity to endure. For humans, _____ is the long-term maintenance of responsibility, which has environmental, economic, and social dimensions, and encompasses the concept of stewardship, the responsible management of resource use. In ecology, _____ describes how biological systems remain diverse and productive over time, a necessary precondition for human well-being.

   a. Sustainability
   b. BEST Education Network
   c. Best of the Web Directory
   d. Biocapacity

3. _____, often referred to as brown coal, or Rosebud coal by Northern Pacific Railroad, is a soft brown fuel with characteristics that put it somewhere between coal and peat. It is considered the lowest rank of coal; it is mined in Greece, Germany, Poland, Serbia, Russia, the United States, India, Australia and many other parts of Europe and it is used almost exclusively as a fuel for steam-electric power generation. Up to 50% of Greece's electricity and 24.6% of Germany's comes from _____ power plants.

   a. Lignite
   b. Ozone-oxygen cycle
   c. Ozone Mapping and Profiler Suite
   d. Ozone-oxygen cycle

4. '_____' systems are a diverse group of air pollution control devices that can be used to remove some particulates and/or gases from industrial exhaust streams. Traditionally, the term '_____' has referred to pollution control devices that use liquid to wash unwanted pollutants from a gas stream. Recently, the term is also used to describe systems that inject a dry reagent or slurry into a dirty exhaust stream to 'wash out' acid gases.

---

a. Selective catalytic reduction
b. Septic drain field
c. Green Bridge
d. Scrubber

5. The _____ are a mountain range rising along the Tennessee-North Carolina border in the southeastern United States. They are a subrange of the Appalachian Mountains, and form part of the Blue Ridge Physiographic Province. The range is sometimes called the Smoky or Smokey Mountains, and the name is commonly shortened to the Smokies.

a. Great Smoky Mountains
b. Jizera Mountains
c. Joyce Kilmer-Slickrock Wilderness
d. King Range Wilderness

1. a
2. a
3. a
4. d
5. a

You can take the complete Chapter Practice Test

for Chapter 14. Air Pollution: Causes, Effects, and Stratospheric Ozone Depletion
on all key terms, persons, places, and concepts.

Online 99 Cents

http://www.epub89.16.20190.14.cram101.com/

Use www.Cram101.com for all your study needs

including Cram101's online interactive problem solving labs in chemistry, statistics, mathematics, and more.

Waste: Solid Waste Generation and Disposal

_____	Landfill
_____	Polystyrene
_____	Detritivore
_____	Invertebrate
_____	Municipal solid waste
_____	Planned obsolescence
_____	Obsolescence
_____	Source reduction
_____	Air pollution
_____	Pollution
_____	Waste management
_____	Leachate
_____	Environmental justice
_____	Water pollution
_____	Bottom ash
_____	Fly ash
_____	Incineration
_____	Waste-to-energy
_____	Hazardous waste

National Priorities List

Superfund

Love Canal

Michael Braungart

Sustainability

Landfill	A landfill site (also known as tip, dump or rubbish dump and historically as a midden) is a site for the disposal of waste materials by burial and is the oldest form of waste treatment. Historically, landfills have been the most common methods of organized waste disposal and remain so in many places around the world.
	Landfills may include internal waste disposal sites (where a producer of waste carries out their own waste disposal at the place of production) as well as sites used by many producers.
Polystyrene	Polystyrene is an aromatic polymer made from the monomer styrene, a liquid hydrocarbon that is manufactured from petroleum by the chemical industry. Polystyrene is one of the most widely used plastics, the scale being several billion kilograms per year.
	Polystyrene is a thermoplastic substance, which is in solid (glassy) state at room temperature, but flows if heated above its glass transition temperature of about 100 °C (for molding or extrusion), and becomes solid again when cooled.
Detritivore	Detritivores, also known as detritophages or detritus feeders or detritus eaters or saprophages, are heterotrophs that obtain nutrients by consuming detritus (decomposing organic matter). By doing so, they contribute to decomposition and the nutrient cycles. They should be distinguished from other decomposers, such as many species of bacteria, fungi and protists, unable to ingest discrete lumps of matter, instead live by absorbing and metabolising on a molecular scale.
Invertebrate	An invertebrate is an animal without a backbone. The group includes 97% of all animal species - all animals except those in the chordate subphylum Vertebrata (fish, amphibians, reptiles, birds, and mammals).
	Invertebrates form a paraphyletic group.

Municipal solid waste	Municipal solid waste commonly known as trash or garbage (US), refuse or rubbish (UK) is a waste type consisting of everyday items that are discard by the public.
	Composition The composition of municipal waste varies greatly from country to country and changes significantly with time.
	In countries which have a developed recycling culture, the waste stream consists mainly of intractable wastes such as plastic film, and un-recyclable packaging.
Planned obsolescence	Planned obsolescence is a licy of planning or designing a product with a limited useful life, so it will become obsolete, that is, unfashionable or no longer functional after a certain period of time. Planned obsolescence has tential benefits for a producer because to obtain continuing use of the product the consumer is under pressure to purchase again, whether from the same manufacturer (a replacement part or a newer model), or from a competitor which might also rely on planned obsolescence.
	In some cases, deliberate deprecation of earlier versions of a technology is used to reduce ongoing suprt costs, especially in the software industry.
Obsolescence	Obsolescence is the state of being which occurs when an object, service or practice is no longer wanted even though it may still be in good working order. Obsolescence frequently occurs because a replacement has become available that is superior in one or more aspects. Obsolete refers to something that is already disused or discarded, or antiquated.
Source reduction	Source reduction refers to any change in the design, manufacture, purchase, or use of materials or products (including packaging) to reduce their amount or toxicity before they become municipal solid waste.
	Synonyms

Pollution Prevention (or P2) and Toxics use reduction are also called source reduction because they address the use of hazardous substances at the source.

Procedures

Source Reduction is achieved through improvements in production and product design, or through Environmentally Preferable Purchasing (EPP).

Air pollution	Air pollution is the introduction of chemicals, particulate matter, or biological materials that cause harm or discomfort to humans or other living organisms, or cause damage to the natural environment or built environment, into the atmosphere. The atmosphere is a complex dynamic natural gaseous system that is essential to support life on planet Earth. Stratospheric ozone depletion due to air pollution has long been recognized as a threat to human health as well as to the Earth's ecosystems.
Pollution	Pollution is the introduction of contaminants into a natural environment that causes instability, disorder, harm or discomfort to the ecosystem i.e. physical systems or living organisms. Pollution can take the form of chemical substances or energy, such as noise, heat or light. Pollutants, the components of pollution, can be either foreign substances/energies or naturally occurring contaminants.
Waste management	Waste management is the collection, transport, processing or disposal, managing and monitoring of waste materials. The term usually relates to materials produced by human activity, and the process is generally undertaken to reduce their effect on health, the environment or aesthetics. Waste management is a distinct practice from resource recovery which focuses on delaying the rate of consumption of natural resources.
Leachate	Leachate is any liquid that, in passing through matter, extracts solutes, suspended solids or any other component of the material through which it has passed.

Leachate is a widely used term in the Environmental sciences where it has the specific meaning of a liquid that has dissolved or entrained environmentally harmful substances which may then enter the environment. It is most commonly used in the context of land-filling of putrescible or industrial waste.

Environmental justice	Environmental justice is 'the fair treatment and meaningful involvement of all people regardless of race, color, sex, national origin, or income with respect to the development, implementation and enforcement of environmental laws, regulations, and policies.' In the words of Bunyan Bryant, 'Environmental justice is served when people can realize their highest potential.'
	Environmental justice emerged as a concept in the United States in the early 1980s; its proponents generally view the environment as encompassing 'where we live, work, and play' (sometimes 'pray' and 'learn' are also included) and seek to redress inequitable distributions of environmental burdens (pollution, industrial facilities, crime, etc).. Root causes of environmental injustices include 'institutionalized racism; the co-modification of land, water, energy and air; unresponsive, unaccountable government policies and regulation; and lack of resources and power in affected communities.'
	Definition The United States Environmental Protection Agency defines as follows:
	'Environmental Justice is the fair treatment and meaningful involvement of all people regardless of race, color, national origin, or income with respect to the development, implementation, and enforcement of environmental laws, regulations, and policies. EPA has this goal for all communities and persons across this Nation.
Water pollution	Water pollution is the contamination of water bodies (e.g. lakes, rivers, oceans, aquifers and groundwater). Water pollution occurs when pollutants are discharged directly or indirectly into water bodies without adequate treatment to remove harmful compounds.
	Water pollution affects plants and organisms living in these bodies of water.

Bottom ash	Bottom ash refers to part of the non-combustible residues of combustion. In an industrial context, it usually refers to coal combustion and comprises traces of combustibles embedded in forming clinkers and sticking to hot side walls of a coal-burning furnace during its operation. The portion of the ash that escapes up the chimney or stack is, however, referred to as fly ash.
Fly ash	Fly ash is one of the residues generated in combustion, and comprises the fine particles that rise with the flue gases. Ash which does not rise is termed bottom ash. In an industrial context, fly ash usually refers to ash produced during combustion of coal.
Incineration	Incineration is a waste treatment process that involves the combustion of organic substances contained in waste materials. Incineration and other high temperature waste treatment systems are described as 'thermal treatment'. Incineration of waste materials converts the waste into ash, flue gas, and heat.
Waste-to-energy	Waste-to-energy or energy-from-waste (EfW) is the process of creating energy in the form of electricity or heat from the incineration of waste source. is a form of energy recovery. Most processes produce electricity directly through combustion, or produce a combustible fuel commodity, such as methane, methanol, ethanol or synthetic fuels.
Hazardous waste	A Hazardous waste is waste that poses substantial or potential threats to public health or the environment. In the United States, the treatment, storage and disposal of hazardous waste is regulated under the Resource Conservation and Recovery Act (RCRA). Hazardous wastes are defined under RCRA in 40 CFR 261 where they are divided into two major categories: characteristic wastes and listed wastes.
National Priorities List	The National Priorities List is the list of hazardous waste sites in the United States eligible for long-term remedial action (cleanup) financed under the federal Superfund program. Environmental Protection Agency (EPA) regulations outline a formal process for assessing hazardous waste sites and placing them on the . The is intended primarily to guide EPA in determining which sites warrant further investigation. The inclusion of a facility in the National Priorities List does not reflect a judgment of its owner or operator or make the owner or operator take any action.

Superfund	Superfund is the common name for the Comprehensive Environmental Response, Compensation, and Liability Act of 1980 (CERCLA), a United States federal law designed to clean up sites contaminated with hazardous substances. Superfund created the Agency for Toxic Substances and Disease Registry (ATSDR), and it provides broad federal authority to clean up releases or threatened releases of hazardous substances that may endanger public health or the environment. The law authorized the Environmental Protection Agency (EPA) to identify parties responsible for contamination of sites and compel the parties to clean up the sites.
Love Canal	Love Canal was a neighborhood in Niagara Falls, New York, located in the white collar LaSalle section of the city. It officially covers 36 square blocks in the far southeastern corner of the city, along 99th Street and Read Avenue. Two bodies of water define the northern and southern boundaries of the neighborhood: Bergholtz Creek to the north and the Niagara River one-quarter mile (400 m) to the south.
Michael Braungart	Michael Braungart is a German chemist who advocates that humans can reduce our negative environmental impact by redesigning industrial production processes. A former Greenpeace activist who once lived in a tree as protest, he is now considered to be a visionary environmental thinker.
	Founder of EPEA International Umweltforschung GmbH in Hamburg, Germany, and co-founder of MBDC McDonough Braungart Design Chemistry in Charlottesville, Virginia, Dr. Braungart is currently a professor of Process Engineering at the University of Applied Sciences in Suderburg (Fachhochschule Nordostniedersachsen), also serving as director of an interdisciplinary materials flow management masters program.
Sustainability	Sustainability is the capacity to endure. For humans, sustainability is the long-term maintenance of responsibility, which has environmental, economic, and social dimensions, and encompasses the concept of stewardship, the responsible management of resource use. In ecology, sustainability describes how biological systems remain diverse and productive over time, a necessary precondition for human well-being.

1.

_____ is a German chemist who advocates that humans can reduce our negative environmental impact by redesigning industrial production processes. A former Greenpeace activist who once lived in a tree as protest, he is now considered to be a visionary environmental thinker.

Founder of EPEA International Umweltforschung GmbH in Hamburg, Germany, and co-founder of MBDC McDonough Braungart Design Chemistry in Charlottesville, Virginia, Dr. Braungart is currently a professor of Process Engineering at the University of Applied Sciences in Suderburg (Fachhochschule Nordostniedersachsen), also serving as director of an interdisciplinary materials flow management masters program.

a. Michael Braungart
b. Minamata disease
c. Mobro 4000
d. Saint John, New Brunswick harbour cleanup

2. A _____ site (also known as tip, dump or rubbish dump and historically as a midden) is a site for the disposal of waste materials by burial and is the oldest form of waste treatment. Historically, _____s have been the most common methods of organized waste disposal and remain so in many places around the world.

_____s may include internal waste disposal sites (where a producer of waste carries out their own waste disposal at the place of production) as well as sites used by many producers.

a. Molten salt oxidation
b. Landfill
c. NERV
d. Priority product

3. _____ is any liquid that, in passing through matter, extracts solutes, suspended solids or any other component of the material through which it has passed.

_____ is a widely used term in the Environmental sciences where it has the specific meaning of a liquid that has dissolved or entrained environmentally harmful substances which may then enter the environment. It is most commonly used in the context of land-filling of putrescible or industrial waste.

    a. Toxicity characteristic leaching procedure
    b. Leachate
    c. Gibbons v. Ogden
    d. Burn pit

4.   _____ is an aromatic polymer made from the monomer styrene, a liquid hydrocarbon that is manufactured from petroleum by the chemical industry. _____ is one of the most widely used plastics, the scale being several billion kilograms per year.

_____ is a thermoplastic substance, which is in solid (glassy) state at room temperature, but flows if heated above its glass transition temperature of about 100 °C (for molding or extrusion), and becomes solid again when cooled.

    a. Juglone
    b. Municipal solid waste
    c. NERV
    d. Polystyrene

5.   _____ is the state of being which occurs when an object, service or practice is no longer wanted even though it may still be in good working order. _____ frequently occurs because a replacement has become available that is superior in one or more aspects. Obsolete refers to something that is already disused or discarded, or antiquated.

    a. Odyssey
    b. Red mud
    c. Sanitary garden
    d. Obsolescence

---

1. a
2. b
3. b
4. d
5. d

You can take the complete Chapter Practice Test

for Chapter 15. Waste: Solid Waste Generation and Disposal
on all key terms, persons, places, and concepts.

Online 99 Cents

http://www.epub89.16.20190.15.cram101.com/

Use www.Cram101.com for all your study needs

including Cram101's online interactive problem solving labs in chemistry, statistics, mathematics, and more.

Human Health and Toxicology: Environmental Sources of Health Risk

	Pollution
	Water pollution
	Air pollution
	Invertebrate
	Transmission
	Pesticide
	Fly ash
	Biodiversity
	Biodiversity hotspot
	Hotspot
	Virus
	Arsenic
	Asbestos
	Carcinogen
	Radon
	Polychlorinated biphenyl
	Allergen
	Endocrine disruptor
	Hormone

Wastewater

Toxin

Epidemiology

Retrospective

Bisphenol A

Routes

Benthic zone

Bioaccumulation

Biomagnification

Groundwater

Trophic level

Environmental hazard

Asbestosis

Sustainability

Efficiency

Pollution	Pollution is the introduction of contaminants into a natural environment that causes instability, disorder, harm or discomfort to the ecosystem i.e. physical systems or living organisms. Pollution can take the form of chemical substances or energy, such as noise, heat or light. Pollutants, the components of pollution, can be either foreign substances/energies or naturally occurring contaminants.
Water pollution	Water pollution is the contamination of water bodies (e.g. lakes, rivers, oceans, aquifers and groundwater). Water pollution occurs when pollutants are discharged directly or indirectly into water bodies without adequate treatment to remove harmful compounds. Water pollution affects plants and organisms living in these bodies of water.
Air pollution	Air pollution is the introduction of chemicals, particulate matter, or biological materials that cause harm or discomfort to humans or other living organisms, or cause damage to the natural environment or built environment, into the atmosphere. The atmosphere is a complex dynamic natural gaseous system that is essential to support life on planet Earth. Stratospheric ozone depletion due to air pollution has long been recognized as a threat to human health as well as to the Earth's ecosystems.
Invertebrate	An invertebrate is an animal without a backbone. The group includes 97% of all animal species - all animals except those in the chordate subphylum Vertebrata (fish, amphibians, reptiles, birds, and mammals). Invertebrates form a paraphyletic group.

Transmission	A machine consists of a power source and a power transmission system, which provides controlled application of the power. Merriam-Webster defines transmission as: an assembly of parts including the speed-changing gears and the propeller shaft by which the power is transmitted from an engine to a live axle. Often transmission refers simply to the gearbox that uses gears and gear trains to provide speed and torque conversions from a rotating power source to another device.
Pesticide	Pesticides are substances or mixture of substances intended for preventing, destroying, repelling or mitigating any pest. A pesticide may be a chemical, biological agent (such as a virus or bacterium), antimicrobial, disinfectant or device used against any pest. Pests include insects, plant pathogens, weeds, molluscs, birds, mammals, fish, nematodes (roundworms), and microbes that destroy property, spread disease or are vectors for disease or cause nuisance.
Fly ash	Fly ash is one of the residues generated in combustion, and comprises the fine particles that rise with the flue gases. Ash which does not rise is termed bottom ash. In an industrial context, fly ash usually refers to ash produced during combustion of coal.
Biodiversity	Biodiversity is the degree of variation of life forms within a given species, ecosystem, biome, or an entire planet. Biodiversity is a measure of the health of ecosystems. Biodiversity is in part a function of climate.
Biodiversity hotspot	A biodiversity hotspot is a biogeographic region with a significant reservoir of biodiversity that is under threat from humans. The concept of biodiversity hotspots was originated by Norman Myers in two articles in 'The Environmentalist' (1988 ' 1990), revised after thorough analysis by Myers and others in 'Hotspots: Earth's Biologically Richest and Most Endangered Terrestrial Ecoregions'. To qualify as a biodiversity hotspot on Myers 2000 edition of the hotspot-map, a region must meet two strict criteria: it must contain at least 0.5% or 1,500 species of vascular plants as endemics, and it has to have lost at least 70% of its primary vegetation.
Hotspot	The places known as hotspots or hot spots in geology are volcanic regions thought to be fed by underlying mantle that is anomalously hot compared with the mantle elsewhere. They may be on, near to, or far from tectonic plate boundaries. There are two hypotheses to explain them.

Chapter 16. Human Health and Toxicology: Environmental Sources of Health Risk

Virus	A virus is a small infectious agent that can replicate only inside the living cells of organisms.
Arsenic	Arsenic is a chemical element with the symbol As, atomic number 33 and relative atomic mass 74.92. Arsenic occurs in many minerals, usually in conjunction with sulfur and metals, and also as a pure elemental crystal. It was first documented by Albertus Magnus in 1250. Arsenic is a metalloid. It can exist in various allotropes, although only the grey form has important use in industry.
Asbestos	Asbestos is a set of six naturally occurring silicate minerals exploited commercially for their desirable physical properties. They all have in common their asbestiform habit, long, (1:20) thin fibrous crystals. The inhalation of asbestos fibers can cause serious illnesses, including malignant lung cancer, mesothelioma (a formerly rare cancer strongly associated with exposure to amphibole asbestos), and asbestosis (a type of pneumoconiosis).
Carcinogen	A carcinogen is any substance, radionuclide or radiation, that is an agent directly involved in causing cancer. This may be due to the ability to damage the genome or to the disruption of cellular metabolic processes. Several radioactive substances are considered carcinogens, but their carcinogenic activity is attributed to the radiation, for example gamma rays and alpha particles, which they emit.
Radon	Radon is a chemical element with the atomic number 86, and is represented by the symbol Rn. It is a radioactive, colorless, odorless, tasteless noble gas, occurring naturally as the decay product of uranium or thorium. Its most stable isotope, ^{222}Rn, has a half-life of 3.8 days.
Polychlorinated biphenyl	A polychlorinated biphenyl is any of the 209 configurations of organochlorides with 2 to 10 chlorine atoms attached to biphenyl, which is a molecule composed of two benzene rings. The chemical formula for a PCB is $C_{12}H_{10-x}Cl_x$. 130 of the 209 different PCB arrangements and orientations are used commercially.
Allergen	An allergen is any substance that can cause an allergy. Technically, an allergen is a non-parasitic antigen capable of stimulating a type-I hypersensitivity reaction in atopic individuals.

	Most humans mount significant Immunoglobulin E (IgE) responses only as a defense against parasitic infections.
Endocrine disruptor	Endocrine disruptors are chemicals that interfere with endocrine (or hormone system) in animals, including humans. These disruptions can cause cancerous tumors, birth defects, and other developmental disorders. Specifically, they are known to cause learning disabilities, severe attention deficit disorder, cognitive and brain development problems, deformations of the body (including limbs); sexual development problems, feminizing of males or masculine effects on females, etc.
Hormone	A hormone is a chemical released by a cell or a gland in one part of the body that sends out messages that affect cells in other parts of the organism. Only a small amount of hormone is required to alter cell metabolism. In essence, it is a chemical messenger that transports a signal from one cell to another.
Wastewater	Waste Water is any water that has been adversely affected in quality by anthropogenic influence. It comprises liquid waste discharged by domestic residences, commercial properties, industry, and/or agriculture and can encompass a wide range of potential contaminants and concentrations. In the most common usage, it refers to the municipal wastewater that contains a broad spectrum of contaminants resulting from the mixing of wastewaters from different sources.
Toxin	A toxin is a poisonous substance produced by living cells or organisms (technically, although humans are living organisms, man-made substances created by artificial processes usually are not considered toxins by this definition). It was the organic chemist Ludwig Brieger (1849-1919) who first used the term 'toxin'.

For a toxic substance not produced by living organisms, 'toxicant' is the more appropriate term, and 'toxics' is an acceptable plural. |

Chapter 16. Human Health and Toxicology: Environmental Sources of Health Risk

Epidemiology	Epidemiology is the study of the distribution and patterns of health-events, health-characteristics and their causes or influences in well-defined populations. It is the cornerstone method of public health research, and helps inform policy decisions and evidence-based medicine by identifying risk factors for disease and targets for preventive medicine. Epidemiologists are involved in the design of studies, collection and statistical analysis of data, and interpretation and dissemination of results (including peer review and occasional systematic review).
Retrospective	Retrospective generally means to take a look back at events that already have taken place. For example, the term is used in medicine, describing a look back at a patient's medical history or lifestyle. Art An exhibition of works from an extended period of an artist's activity.
Bisphenol A	Bisphenol A is an organic compound with the chemical formula $(CH_3)_2C(C_6H_4OH)_2$. It is a colourless solid that is soluble in organic solvents, but poorly soluble in water. Having two phenol functional groups, it is used to make polycarbonate polymers and epoxy resins, along with other materials used to make plastics.
Routes	Routes is a commune in the Seine-Maritime department in the Haute-Normandie region in northern France. A small farming village in the Pays de Caux, some 33 miles (53 km) northeast of Le Havre, at the junction of the D88 and D420 roads. Population Places of interest • The church of St. Martin-et-Notre-Dame, dating from the eighteenth century.
Benthic zone	The benthic zone is the ecological region at the lowest level of a body of water such as an ocean or a lake, including the sediment surface and some sub-surface layers. Organisms living in this zone are called benthos. They generally live in close relationship with the substrate bottom; many such organisms are permanently attached to the bottom.

Bioaccumulation	Bioaccumulation refers to the accumulation of substances, such as pesticides, or other organic chemicals in an organism. Bioaccumulation occurs when an organism absorbs a toxic substance at a rate greater than that at which the substance is lost. Thus, the longer the biological half-life of the substance the greater the risk of chronic poisoning, even if environmental levels of the toxin are not very high.
Biomagnification	Biomagnification, is the increase in concentration of a substance that occurs in a food chain as a consequence of: • Persistence (can't be broken down by environmental processes) • Food chain energetics • Low (or nonexistent) rate of internal degradation/excretion of the substance (often due to water-insolubility) The following is an example showing how biomagnification takes place in nature: An anchovy eats zooplankton that have tiny amounts of mercury that the zooplankton has picked up from the water throughout the anchovie's lifespan. A tuna eats many of these anchovies over its life, accumulating the mercury in each of those anchovies into its body. If the mercury stunts the growth of the anchovies, that tuna is required to eat more little fish to stay alive.
Groundwater	Groundwater is water located beneath the ground surface in soil pore spaces and in the fractures of rock formations. A unit of rock or an unconsolidated deposit is called an aquifer when it can yield a usable quantity of water. The depth at which soil pore spaces or fractures and voids in rock become completely saturated with water is called the water table.
Trophic level	The trophic level of an organism is the position it occupies in a food chain. A food chain represents a succession of organisms that eat another organism and are, in turn, eaten themselves. The number of steps an organism is from the start of the chain is a measure of its trophic level.

Environmental hazard	'Environmental hazard' is a generic term for any situation or state of events which poses a threat to the surrounding natural environment and adversely affect people's health. This term incorporates topics like pollution and natural disasters such as storms and earthquakes. Hazards can be categorized in five types: 1. Chemical 2. Physical 3. Mechanical 4. Biological 5. Psychosocial Examples - Allergens - Anthrax - Antibiotic agents in animals destined for human consumption - Arbovirus - Arsenic - a contaminant of fresh water sources (water wells) - Asbestos - carcinogenic - Avian influenza - Bovine spongiform encephalopathy (BSE) - Carcinogens - Cholera - Cosmic rays - DDT - dioxins - Drought - Dysentery - Electromagnetic fields - Endocrine disruptors - Epidemics - E-waste - Explosive material - Floods - Food poisoning - Fungicides - Furans - Haloalkanes - Heavy metals - Herbicides - Hormones in animals destined for human consumption - Lead in paint - Light pollution - Lighting

---

Asbestosis	Asbestosis is a chronic inflammatory and fibrotic medical condition affecting the parenchymal tissue of the lungs caused by the inhalation and retention of asbestos fibers. It usually occurs after high intensity and/or long-term exposure to asbestos (particularly in those individuals working on the production or end-use of products containing asbestos) and is therefore regarded as an occupational lung disease. People with extensive occupational exposure to the mining, manufacturing, handling or removal of asbestos are at risk of developing asbestosis.
Sustainability	Sustainability is the capacity to endure. For humans, sustainability is the long-term maintenance of responsibility, which has environmental, economic, and social dimensions, and encompasses the concept of stewardship, the responsible management of resource use. In ecology, sustainability describes how biological systems remain diverse and productive over time, a necessary precondition for human well-being.
Efficiency	Efficiency in general describes the extent to which time or effort is well used for the intended task or purpose. It is often used with the specific purpose of relaying the capability of a specific application of effort to produce a specific outcome effectively with a minimum amount or quantity of waste, expense, or unnecessary effort. 'Efficiency' has widely varying meanings in different disciplines.

1.

_____ is a chemical element with the symbol As, atomic number 33 and relative atomic mass 74.92. _____ occurs in many minerals, usually in conjunction with sulfur and metals, and also as a pure elemental crystal. It was first documented by Albertus Magnus in 1250. _____ is a metalloid. It can exist in various allotropes, although only the grey form has important use in industry.

   a. Atrazine
   b. Arsenic
   c. Iprodione
   d. Buffer zone

2. A _____ is a small infectious agent that can replicate only inside the living cells of organisms.

   a. Alfonso Jordan
   b. Bonn Convention
   c. Virus
   d. Buffer zone

3. _____ is one of the residues generated in combustion, and comprises the fine particles that rise with the flue gases. Ash which does not rise is termed bottom ash. In an industrial context, _____ usually refers to ash produced during combustion of coal.

   a. Food waste
   b. Great Stink
   c. Green waste
   d. Fly ash

4. A _____ is a biogeographic region with a significant reservoir of biodiversity that is under threat from humans.

The concept of _____s was originated by Norman Myers in two articles in 'The Environmentalist' (1988 ' 1990), revised after thorough analysis by Myers and others in 'Hotspots: Earth's Biologically Richest and Most Endangered Terrestrial Ecoregions'.

To qualify as a _____ on Myers 2000 edition of the hotspot-map, a region must meet two strict criteria: it must contain at least 0.5% or 1,500 species of vascular plants as endemics, and it has to have lost at least 70% of its primary vegetation.

a. Biodiversity Indicators Partnership
b. Bonn Convention
c. Breeding season
d. Biodiversity hotspot

5. _____s are chemicals that interfere with endocrine (or hormone system) in animals, including humans. These disruptions can cause cancerous tumors, birth defects, and other developmental disorders. Specifically, they are known to cause learning disabilities, severe attention deficit disorder, cognitive and brain development problems, deformations of the body (including limbs); sexual development problems, feminizing of males or masculine effects on females, etc.

a. Endocrine disruptor
b. Alachlor
c. Aldrin
d. Arsenic

---

1. b
2. c
3. d
4. d
5. a

You can take the complete Chapter Practice Test

for Chapter 16. Human Health and Toxicology: Environmental Sources of Health Risk
on all key terms, persons, places, and concepts.

Online 99 Cents

http://www.epub89.16.20190.16.cram101.com/

Use www.Cram101.com for all your study needs

including Cram101's online interactive problem solving labs in chemistry, statistics, mathematics, and more.

Conservation of Biodiversity: Protection of Earth`s Species and Ecosystems

National park

Biodiversity

Biodiversity hotspot

Hotspot

Extinction

Inbreeding

Inbreeding depression

Air pollution

Pollution

Green Revolution

Species diversity

Deforestation

Habitat

Invertebrate

Millennium Ecosystem Assessment

Sustainable development

Global change

Coral reef

Grassland

	Wetland
	Convection
	Ecosystem services
	Invasive species
	Sewage
	Sewage treatment
	Dodo
	Atmosphere
	Water pollution
	Endangered Species Act
	Island biogeography
	Biogeography
	Theory
	Biosphere
	Habitat corridor
	Debt-for-nature swap
	Sustainability
	Thomas Garnett

National park	A national park is a reserve of natural, semi-natural, or developed land that a sovereign state declares or owns. Although individual nations designate their own national parks differently , an international organization, the International Union for Conservation of Nature (IUCN), and its World Commission on Protected Areas, has defined National Parks as its category II type of protected areas.
Biodiversity	Biodiversity is the degree of variation of life forms within a given species, ecosystem, biome, or an entire planet. Biodiversity is a measure of the health of ecosystems. Biodiversity is in part a function of climate.
Biodiversity hotspot	A biodiversity hotspot is a biogeographic region with a significant reservoir of biodiversity that is under threat from humans.
	The concept of biodiversity hotspots was originated by Norman Myers in two articles in 'The Environmentalist' (1988 ' 1990), revised after thorough analysis by Myers and others in 'Hotspots: Earth's Biologically Richest and Most Endangered Terrestrial Ecoregions'.
	To qualify as a biodiversity hotspot on Myers 2000 edition of the hotspot-map, a region must meet two strict criteria: it must contain at least 0.5% or 1,500 species of vascular plants as endemics, and it has to have lost at least 70% of its primary vegetation.
Hotspot	The places known as hotspots or hot spots in geology are volcanic regions thought to be fed by underlying mantle that is anomalously hot compared with the mantle elsewhere. They may be on, near to, or far from tectonic plate boundaries. There are two hypotheses to explain them.
Extinction	In biology and ecology, extinction is the end of an organism or of a group of organisms (taxon), normally a species. The moment of extinction is generally considered to be the death of the last individual of the species, although the capacity to breed and recover may have been lost before this point. Because a species' potential range may be very large, determining this moment is difficult, and is usually done retrospectively.
Inbreeding	Inbreeding is the reproduction from the mating of two genetically related parents, which can increase the chances of offspring being affected by recessive or deleterious traits. This generally leads to a decreased fitness of a population, which is called inbreeding depression. Deleterious alleles causing inbreeding depression can subsequently be removed through culling, which is also known as genetic purging.

Inbreeding depression	Inbreeding depression is the reduced fitness in a given population as a result of breeding of related individuals. It is often the result of a population bottleneck. In general, the higher the genetic variation within a breeding population, the less likely it is to suffer from inbreeding depression.
Air pollution	Air pollution is the introduction of chemicals, particulate matter, or biological materials that cause harm or discomfort to humans or other living organisms, or cause damage to the natural environment or built environment, into the atmosphere. The atmosphere is a complex dynamic natural gaseous system that is essential to support life on planet Earth. Stratospheric ozone depletion due to air pollution has long been recognized as a threat to human health as well as to the Earth's ecosystems.
Pollution	Pollution is the introduction of contaminants into a natural environment that causes instability, disorder, harm or discomfort to the ecosystem i.e. physical systems or living organisms. Pollution can take the form of chemical substances or energy, such as noise, heat or light. Pollutants, the components of pollution, can be either foreign substances/energies or naturally occurring contaminants.
Green Revolution	Green Revolution refers to a series of research, development, and technology transfer initiatives, occurring between the 1940s and the late 1970s, that increased agriculture production around the world, beginning most markedly in the late 1960s. The initiatives, led by Norman Borlaug, the 'Father of the Green Revolution' credited with saving over a billion people from starvation, involved the development of high-yielding varieties of cereal grains, expansion of irrigation infrastructure, modernization of management techniques, distribution of hybridized seeds, synthetic fertilizers, and pesticides to farmers. The term 'Green Revolution' was first used in 1968 by former United States Agency for International Development (USAID) director William Gaud, who noted the spread of the new technologies and said,

These and other developments in the field of agriculture contain the makings of a new revolution.

Species diversity	Species diversity is the effective number of different species that are represented in a collection of individuals (a dataset). The effective number of species refers to the number of equally-abundant species needed to obtain the same mean proportional species abundance as that observed in the dataset of interest (where all species may not be equally abundant). Species diversity consists of two components, species richness and species evenness.
Deforestation	Deforestation is the removal of a forest or stand of trees where the land is thereafter converted to a nonforest use. Examples of deforestation include conversion of forestland to farms, ranches, or urban use. The term deforestation is often misused to describe any activity where all trees in an area are removed.
Habitat	A habitat is an ecological or environmental area that is inhabited by a particular species of animal, plant or other type of organism. It is the natural environment in which an organism lives, or the physical environment that surrounds (influences and is utilized by) a species population. Definition The term 'population' is preferred to 'organism' because, while it is possible to describe the habitat of a single black bear, it is also possible that one may not find any particular or individual bear but the grouping of bears that constitute a breeding population and occupy a certain biogeographical area.
Invertebrate	An invertebrate is an animal without a backbone. The group includes 97% of all animal species - all animals except those in the chordate subphylum Vertebrata (fish, amphibians, reptiles, birds, and mammals).

Invertebrates form a paraphyletic group.

Millennium Ecosystem Assessment	The Millennium Ecosystem Assessment, released in 2005, is an international synthesis by over 1000 of the world's leading biological scientists that analyses the state of the Earth's ecosystems and provides summaries and guidelines for decision-makers. It concludes that human activity is having a significant and escalating impact on the biodiversity of world ecosystems, reducing both their resilience and biocapacity. The report refers to natural systems as humanity's 'life-support system', providing essential 'ecosystem services'.
Sustainable development	Sustainable development is a pattern of growth in which resource use aims to meet human needs while preserving the environment so that these needs can be met not only in the present, but also for generations to come (sometimes taught as ELF-Environment, Local people, Future). The term sustainable development was used by the Brundtland Commission which coined what has become the most often-quoted definition of sustainable development as development that 'meets the needs of the present without compromising the ability of future generations to meet their own needs.' Sustainable development ties together concern for the carrying capacity of natural systems with the social challenges facing humanity. As early as the 1970s 'sustainability' was employed to describe an economy 'in equilibrium with basic ecological support systems.' Ecologists have pointed to The Limits to Growth, and presented the alternative of a 'steady state economy' in order to address environmental concerns.
Global change	Global change refers to planetary-scale changes in the Earth system. The system consists of the land, oceans, atmosphere, poles, life, the planet's natural cycles and deep Earth processes. These constituent parts influence one another.
Coral reef	Coral reefs are underwater structures made from calcium carbonate secreted by corals. Corals are colonies of tiny living animals found in marine waters that contain few nutrients. Most coral reefs are built from stony corals, which in turn consist of polyps that cluster in groups.

Grassland	Grasslands are areas where the vegetation is dominated by grasses (Poaceae) and other herbaceous (non-woody) plants (forbs). However, sedge (Cyperaceae) and rush (Juncaceae) families can also be found. Grasslands occur naturally on all continents except Antarctica. In temperate latitudes, such as northwestern Europe and the Great Plains and California in North America, native grasslands are dominated by perennial bunch grass species, whereas in warmer climates annual species form a greater component of the vegetation.
Wetland	A wetland is a land area that is saturated with water, either permanently or seasonally, such that it takes on characteristics that distinguish it as a distinct ecosystem. The primary factor that distinguishes wetlands is the characteristic vegetation that is adapted to its unique soil conditions: Wetlands are made up primarily of hydric soil, which supports aquatic plants. The water found in wetlands can be saltwater, freshwater, or brackish.
Convection	Convection is the concerted, collective movement of ensembles of molecules within fluids (i.e. liquids, gases) and rheids. Convection of mass cannot take place in solids, since neither bulk current flows nor significant diffusion can take place in solids. Diffusion of heat can take place in solids, but is referred to separately in that case as heat conduction.
Ecosystem services	Humankind benefits from a multitude of resources and processes that are supplied by natural ecosystems. Collectively, these benefits are known as ecosystem services and include products like clean drinking water and processes such as the decomposition of wastes. While scientists and environmentalists have discussed ecosystem services for decades, these services were popularized and their definitions formalized by the United Nations 2004 Millennium Ecosystem Assessment (MA), a four-year study involving more than 1,300 scientists worldwide.

Invasive species	Invasive species, a nomenclature term and categorization phrase used for flora and fauna, and for specific restoration-preservation processes in native habitats, with several definitions. • The first definition, the most used, applies to introduced species (also called 'non-indigenous' or 'non-native') that adversely affect the habitats and bioregions they invade economically, environmentally, and/or ecologically. Such invasive species may be either plants or animals and may drupt by dominating a region, wilderness areas, particular habitats, or wildland-urban interface land from loss of natural controls (such as predators or herbivores).
Sewage	Sewage is water-carried waste, in solution or suspension, that is intended to be removed from a community. Also known as wastewater, it is more than 99% water and is characterized by volume or rate of flow, physical condition, chemical constituents and the bacteriological organisms that it contains. In loose American English usage, the terms 'sewage' and 'sewerage' are sometimes interchanged.
Sewage treatment	Sewage treatment, is the process of removing contaminants from waewater and household sewage, both runoff (effluents) and domeic. It includes physical, chemical, and biological processes to remove physical, chemical and biological contaminants. Its objective is to produce an environmentally-safe fluid wae ream (or treated effluent) and a solid wae (or treated sludge) suitable for disposal or reuse (usually as farm fertilizer).
Dodo	The dodo was a flightless bird endemic to the Indian Ocean island of Mauritius. It stood about a metre (3.3 feet) tall, weighing about 20 kilograms (44 lb). The dodo lost the power of flight because food was abundant and mammalian predators were absent on Mauritius.
Atmosphere	The standard atmosphere (symbol: atm) is an international reference pressure defined as 101325 Pa and formerly used as unit of pressure. For practical purposes it has been replaced by the bar which is 10^5 Pa. The difference of about 1% is not significant for many applications, and is within the error range of common pressure gges.
Water pollution	Water pollution is the contamination of water bodies (e.g. lakes, rivers, oceans, aquifers and groundwater). Water pollution occurs when pollutants are discharged directly or indirectly into water bodies without adequate treatment to remove harmful compounds.

Water pollution affects plants and organisms living in these bodies of water.

Endangered Species Act

The Endangered Species Act of 1973 (Endangered Species Act; 7 U.S.C. § 136, 16 U.S.C. § 1531 et seq). is one of the dozens of United States environmental laws passed in the 1970s. Signed into law by President Richard Nixon on December 28, 1973, it was designed to protect critically imperiled species from extinction as a 'consequence of economic growth and development untempered by adequate concern and conservation.'

The Act is administered by two federal agencies, the United States Fish and Wildlife Service (FWS) and the National Oceanic and Atmospheric Administration (NOAA).

Island biogeography

Island biogeography is a field within biogeography that attempts to establish and explain the factors that affect the species richness of natural communities. The theory was developed to explain species richness of actual islands. It has since been extended to mountains surrounded by deserts, lakes surrounded by dry land, fragmented forest and even natural habitats surrounded by human-altered landscapes.

Biogeography

Biogeography is the study of the distribution of species (biology), organisms, and ecosystems in space and through geological time. Organisms and biological communities vary in a highly regular fashion along geographic gradients of latitude, elevation, isolation and habitat area.

Knowledge of spatial variation in the numbers and types of organisms is as vital to us today as it was to our early human ancestors, as we adapt to heterogeneous but geographically predictable environments.

Theory

The English word theory was derived from a technical term in Ancient Greek philosophy. The word theoria, θεωρ?α, meant 'a looking at, viewing, beholding', and referring to contemplation or speculation, as opposed to action. Theory is especially often contrasted to 'practice' a Greek term for 'doing', which is opposed to theory because theory involved no doing apart from itself.

Biosphere	The biosphere is the global sum of all ecosystems. It can also be called the zone of life on Earth, a closed (apart from solar and cosmic radiation) and self-regulating system. From the broadest biophysiological point of view, the biosphere is the global ecological system integrating all living beings and their relationships, including their interaction with the elements of the lithosphere, hydrosphere and atmosphere.
Habitat corridor	A habitat corridor is a strip of land that aids in the movement of species between disconnected areas of their natural habitat. An animal's natural habitat would typically include a number of areas necessary to thrive, such as wetlands, burrowing sites, food, and breeding grounds. Urbanization can split up such areas, causing animals to lose both their natural habitat and the ability to move between regions to use all of the resources they need to survive.
Debt-for-nature swap	Debt-for-nature swaps are financial transactions in which a portion of a developing nation's foreign debt is forgiven in exchange for local investments in environmental conservation measures. History The concept of debt-for-nature swaps was first conceived by Thomas Lovejoy of the World Wildlife Fund in 1984 as an opportunity to deal with the problems of developing-nation indebtedness and its consequent deleterious effect on the environment. In the wake of the Latin American debt crisis that resulted in steep reductions to the environmental conservation ability of highly-indebted nations, Lovejoy suggested that ameliorating debt and promoting conservation could be done at the same time.
Sustainability	Sustainability is the capacity to endure. For humans, sustainability is the long-term maintenance of responsibility, which has environmental, economic, and social dimensions, and encompasses the concept of stewardship, the responsible management of resource use. In ecology, sustainability describes how biological systems remain diverse and productive over time, a necessary precondition for human well-being.
Thomas Garnett	Thomas Garnett (1766-1802) was an English physician and natural philosopher. Life He was born 21 April 1766 at Casterton in Westmoreland, where his father had a small landed property. After attending a local school he was at the age of fifteen articled at his own request to the celebrated John Dawson (surgeon) of Sedbergh, Yorkshire, surgeon and mathematician.

---

1. _____ is the reduced fitness in a given population as a result of breeding of related individuals. It is often the result of a population bottleneck. In general, the higher the genetic variation within a breeding population, the less likely it is to suffer from _____.

   a. Inbreeding depression
   b. Bonn Convention
   c. Breeding season
   d. Buffer zone

2. A _____ is a biogeographic region with a significant reservoir of biodiversity that is under threat from humans.

   The concept of _____s was originated by Norman Myers in two articles in 'The Environmentalist' (1988 ' 1990), revised after thorough analysis by Myers and others in 'Hotspots: Earth's Biologically Richest and Most Endangered Terrestrial Ecoregions'.

   To qualify as a _____ on Myers 2000 edition of the hotspot-map, a region must meet two strict criteria: it must contain at least 0.5% or 1,500 species of vascular plants as endemics, and it has to have lost at least 70% of its primary vegetation.

   a. Biodiversity Indicators Partnership
   b. Bonn Convention
   c. Biodiversity hotspot
   d. Buffer zone

3. The _____ of 1973 (_____; 7 U.S.C. § 136, 16 U.S.C. § 1531 et seq). is one of the dozens of United States environmental laws passed in the 1970s. Signed into law by President Richard Nixon on December 28, 1973, it was designed to protect critically imperiled species from extinction as a 'consequence of economic growth and development untempered by adequate concern and conservation.'

   The Act is administered by two federal agencies, the United States Fish and Wildlife Service (FWS) and the National Oceanic and Atmospheric Administration (NOAA).

---

a. Endangered Species Act
b. Bioassay
c. Bioretention
d. Biosurvey

4. A _____ is an ecological or environmental area that is inhabited by a particular species of animal, plant or other type of organism. It is the natural environment in which an organism lives, or the physical environment that surrounds (influences and is utilized by) a species population.

Definition
The term 'population' is preferred to 'organism' because, while it is possible to describe the _____ of a single black bear, it is also possible that one may not find any particular or individual bear but the grouping of bears that constitute a breeding population and occupy a certain biogeographical area.

a. Habitat Conservation Plan
b. Habitat
c. Landscape limnology
d. Levels of organization

5. _____ is a field within biogeography that attempts to establish and explain the factors that affect the species richness of natural communities. The theory was developed to explain species richness of actual islands. It has since been extended to mountains surrounded by deserts, lakes surrounded by dry land, fragmented forest and even natural habitats surrounded by human-altered landscapes.

a. Odyssey
b. Bioassay
c. Island biogeography
d. Biosurvey

1. a
2. c
3. a
4. b
5. c

You can take the complete Chapter Practice Test

for Chapter 17. Conservation of Biodiversity: Protection of Earth`s Species and Ecosystems
on all key terms, persons, places, and concepts.

Online 99 Cents

http://www.epub89.16.20190.17.cram101.com/

Use www.Cram101.com for all your study needs

including Cram101's online interactive problem solving labs in chemistry, statistics, mathematics, and more.

Global Change: Climate Alteration and Global Warming

Ice cap

Global warming

Bioaccumulation

Global change

Air pollution

Climate change

Pollution

Greenhouse effect

Radiation

Ozone

Ozone layer

Carbon dioxide

Chlorofluorocarbon

Methane

Greenhouse gas

Potential

Invertebrate

Mount Pinatubo

Denitrification

Fossil

Green Revolution

Carbon cycle

Environmental justice

Montreal Protocol

Charles David Keeling

Biome

Ice core

Carrying capacity

Demographic transition

Isotope

Climate model

Decomposer

Negative feedback

Permafrost

Positive feedback

Glacier

Sea level

Heat wave

Chapter 18. Global Change: Climate Alteration and Global Warming

_____ | Hurricane _____

_____ | Invasive species _____

_____ | Coral bleaching _____

_____ | Fly ash _____

_____ | Growing season _____

_____ | Gulf Stream _____

_____ | Thermohaline circulation _____

_____ | Biodiversity _____

_____ | Biodiversity hotspot _____

_____ | Hotspot _____

_____ | Scientific method _____

_____ | Water pollution _____

_____ | Uncertainty _____

_____ | Kyoto Protocol _____

_____ | Carbon sequestration _____

_____ | Efficiency _____

_____ | Sustainability _____

CHAPTER HIGHLIGHTS: KEY TERMS, PEOPLE, PLACES, CONCEPTS
Chapter 18. Global Change: Climate Alteration and Global Warming

267

Ice cap	An ice cap is an ice mass that covers less than 50 000 km² of land area (usually covering a highland area). Masses of ice covering more than 50 000 km² are termed an ice sheet. Ice caps are not constrained by topographical features (i.e., they will lie over the top of mountains) but their dome is usually centred on the highest point of a massif.
Global warming	Global warming refers to the rising average temperature of Earth's atmosphere and oceans, which began to increase in the late 19th century and is projected to continue rising. Since the early 20th century, Earth's average surface temperature has increased by about 0.8 °C (1.4 °F), with about two thirds of the increase occurring since 1980. Warming of the climate system is unequivocal, and scientists are more than 90% certain that most of it is caused by increasing concentrations of greenhouse gases produced by human activities such as deforestation and the burning of fossil fuels. These findings are recognized by the national science academies of all major industrialized nations.[A] Climate model projections are summarized in the 2007 Fourth Assessment Report (AR4) by the Intergovernmental Panel on Climate Change (IPCC).
Bioaccumulation	Bioaccumulation refers to the accumulation of substances, such as pesticides, or other organic chemicals in an organism. Bioaccumulation occurs when an organism absorbs a toxic substance at a rate greater than that at which the substance is lost. Thus, the longer the biological half-life of the substance the greater the risk of chronic poisoning, even if environmental levels of the toxin are not very high.
Global change	Global change refers to planetary-scale changes in the Earth system. The system consists of the land, oceans, atmosphere, poles, life, the planet's natural cycles and deep Earth processes. These constituent parts influence one another.
Air pollution	Air pollution is the introduction of chemicals, particulate matter, or biological materials that cause harm or discomfort to humans or other living organisms, or cause damage to the natural environment or built environment, into the atmosphere.

The atmosphere is a complex dynamic natural gaseous system that is essential to support life on planet Earth. Stratospheric ozone depletion due to air pollution has long been recognized as a threat to human health as well as to the Earth's ecosystems.

Climate change

Climate change is a significant and lasting change in the statistical distribution of weather patterns over periods ranging from decades to millions of years. It may be a change in average weather conditions or the distribution of events around that average (e.g., more or fewer extreme weather events). Climate change may be limited to a specific region or may our across the whole Earth.

Pollution

Pollution is the introduction of contaminants into a natural environment that causes instability, disorder, harm or discomfort to the ecosystem i.e. physical systems or living organisms. Pollution can take the form of chemical substances or energy, such as noise, heat or light. Pollutants, the components of pollution, can be either foreign substances/energies or naturally occurring contaminants.

Greenhouse effect

The greenhouse effect is a process by which thermal radiation from a planetary surface is absorbed by atmospheric greenhouse gases, and is re-radiated in all directions. Since part of this re-radiation is back towards the surface, energy is transferred to the surface and the lower atmosphere. As a result, the avera surface temperature is higher than it would be if direct heating by solar radiation were the only warming mechanism.

Radiation

In physics, radiation is a process in which energetic particles or energetic waves travel through a medium or space. Two types of radiation are commonly differentiated in the way they interact with normal chemical matter: ionizing and non-ionizing radiation. The word radiation is often colloquially used in reference to ionizing radiation but the term radiation may correctly also refer to non-ionizing radiation.

Ozone

Ozone or trioxygen, is a triatomic molecule, consisting of three oxygen atoms. It is an allotrope of oxygen that is much less stable than the diatomic allotrope (O_2). Ozone in the lower atmosphere is an air pollutant with harmful effects on the respiratory systems of animals and will burn sensitive plants; however, the ozone layer in the upper atmosphere is beneficial, preventing damaging ultraviolet light from reaching the Earth's surface.

Ozone layer	The ozone layer is a layer in Earth's atmosphere which contains relatively high concentrations of ozone (O_3). This layer absorbs 97-99% of the Sun's high frequency ultraviet light, which potentially damages the life forms on Earth. It is mainly located in the lower portion of the stratosphere from approximately 20 to 30 kilometres (12 to 19 mi) above Earth, though the thickness varies seasonally and geographically.
Carbon dioxide	Carbon dioxide is a naturally occurring chemical compound composed of two oxygen atoms covalently bonded to a single carbon atom. It is a gas at standard temperature and pressure and exists in Earth's atmosphere in this state, as a trace gas at a concentration of 0.039% by volume. As part of the carbon cycle known as photosynthesis, plants, algae, and cyanobacteria absorb carbon dioxide, light, and water to produce carbohydrate energy for themselves and oxygen as a waste product.
Chlorofluorocarbon	A chlorofluorocarbon is an organic compound that contains carbon, chlorine, and fluorine, produced as a volatile derivative of methane and ethane. A common subclass are the hydrochlorofluorocarbons (HCFCs), which contain hydrogen, as well. They are also commonly known by the DuPont trade name Freon.
Methane	Appendix: extraterrestrial methane Methane has been detected or is believed to exist in several locations of the solar system. In most cases, it is believed to have been created by abiotic processes. Possible exceptions are Mars and Titan.
Greenhouse gas	A greenhouse gas is a gas in an atmosphere that absorbs and emits radiation within the thermal infrared range. This process is the fundamental cause of the greenhouse effect. The primary greenhouse gases in the Earth's atmosphere are water vapor, carbon dioxide, methane, nitrous oxide, and ozone.

Potential	<ul><li>In linguistics, the potential mood</li><li>The mathematical study of potentials is known as potential theory; it is the study of harmonic functions on manifolds. This mathematical formulation arises from the fact that, in physics, the scalar potential is irrotational, and thus has a vanishing Laplacian -- the very definition of a harmonic function.</li><li>In physics, a potential may refer to the scalar potential or to the vector potential. In either case, it is a field defined in space, from which many important physical properties may be derived.</li></ul>
Invertebrate	An invertebrate is an animal without a backbone. The group includes 97% of all animal species - all animals except those in the chordate subphylum Vertebrata (fish, amphibians, reptiles, birds, and mammals). Invertebrates form a paraphyletic group.
Mount Pinatubo	Mount Pinatubo is an active stratovolcano located on the island of Luzon, at the intersection of the borders of the Philippine provinces of Zambales, Tarlac, and Pampanga. It is located in the Tri-Cabusilan Mountain range separating the west coast of Luzon from the central plains, and is 42 km (26 mi) west of the dormant and more prominent Mount Arayat, occasionally mistaken for Pinatubo. Ancestral Pinatubo was a stratovolcano made of andesite and dacite.
Denitrification	Denitrification is a microbially facilitated process of nitrate reduction that may ultimately produce molecular nitrogen (N_2) through a series of intermediate gaseous nitrogen oxide products. This respiratory process reduces oxidized forms of nitrogen in response to the oxidation of an electron donor such as organic matter. The preferred nitrogen electron acceptors in order of most to least thermodynamically favorable include nitrate (NO_3^-), nitrite (NO_2^-), nitric oxide (NO), and nitrous oxide (N_2O).

Fossil	Fossils are the preserved remains or traces of animals (also known as zoolites), plants, and other organisms from the remote past. The totality of fossils, both discovered and undiscovered, and their placement in fossiliferous (fossil-containing) rock formations and sedimentary layers (strata) is known as the fossil record.
	The study of fossils across geological time, how they were formed, and the evolutionary relationships between taxa (phylogeny) are some of the most important functions of the science of paleontology.
Green Revolution	Green Revolution refers to a series of research, development, and technology transfer initiatives, occurring between the 1940s and the late 1970s, that increased agriculture production around the world, beginning most markedly in the late 1960s.
	The initiatives, led by Norman Borlaug, the 'Father of the Green Revolution' credited with saving over a billion people from starvation, involved the development of high-yielding varieties of cereal grains, expansion of irrigation infrastructure, modernization of management techniques, distribution of hybridized seeds, synthetic fertilizers, and pesticides to farmers.
	The term 'Green Revolution' was first used in 1968 by former United States Agency for International Development (USAID) director William Gaud, who noted the spread of the new technologies and said,
	These and other developments in the field of agriculture contain the makings of a new revolution.
Carbon cycle	The carbon cycle is the biogeochemical cycle by which carbon is exchanged among the biosphere, pedosphere, geosphere, hydrosphere, and atmosphere of the Earth. It is one of the most important cycles of the earth and allows for carbon to be recycled and reused throughout the biosphere and all of its organisms.

The carbon cycle was initially discovered by Joseph Priestley and Antoine Lavoisier, and popularized by Humphry Davy.

Environmental justice	Environmental justice is 'the fair treatment and meaningful involvement of all people regardless of race, color, sex, national origin, or income with respect to the development, implementation and enforcement of environmental laws, regulations, and policies.' In the words of Bunyan Bryant, 'Environmental justice is served when people can realize their highest potential.' Environmental justice emerged as a concept in the United States in the early 1980s; its proponents generally view the environment as encompassing 'where we live, work, and play' (sometimes 'pray' and 'learn' are also included) and seek to redress inequitable distributions of environmental burdens (pollution, industrial facilities, crime, etc).. Root causes of environmental injustices include 'institutionalized racism; the co-modification of land, water, energy and air; unresponsive, unaccountable government policies and regulation; and lack of resources and power in affected communities.' Definition The United States Environmental Protection Agency defines as follows: 'Environmental Justice is the fair treatment and meaningful involvement of all people regardless of race, color, national origin, or income with respect to the development, implementation, and enforcement of environmental laws, regulations, and policies. EPA has this goal for all communities and persons across this Nation.

Montreal Protocol	The Montreal Protocol on Substances That Deplete the Ozone Layer (a protocol to the Vienna Convention for the Protection of the Ozone Layer) is an international treaty designed to protect the ozone layer by phasing out the production of numerous substances believed to be responsible for ozone depletion. The treaty was opened for signature on September 16, 1987, and entered into force on January 1, 1989, followed by a first meeting in Helsinki, May 1989. Since then, it has undergone seven revisions, in 1990 (London), 1991 (Nairobi), 1992 (Copenhagen), 1993 (Bangkok), 1995 (Vienna), 1997 (Montreal), and 1999 (Beijing). It is believed that if the international agreement is adhered to, the ozone layer is expected to recover by 2050. Due to its widespread adoption and implementation it has been hailed as an example of exceptional international co-operation, with Kofi Annan quoted as saying that 'perhaps the single most successful international agreement to date has been the Montreal Protocol'.
Charles David Keeling	Charles David Keeling was an American scientist whose recording of carbon dioxide at the Mauna Loa Observatory first alerted the world to the possibility of anthropogenic contribution to the 'greenhouse effect' and global warming. The Keeling Curve measures the progressive buildup of carbon dioxide, a greenhouse gas, in the atmosphere.
Biome	Biomes are climatically and geographically defined as similar climatic conditions on the Earth, such as communities of plants, animals, and soil organisms, and are often referred to as ecosystems. Some parts of the earth have more or less the same kind of abiotic and biotic factors spread over a large area, creating a typical ecosystem over that area. Such major ecosystems are termed as biomes.
Ice core	An ice core is a core sample that is typically removed from an ice sheet, most commonly from the polar ice caps of Antarctica, Greenland or from high mountain glaciers elsewhere. As the ice forms from the incremental build up of annual layers of snow, lower layers are older than upper, and an ice core contains ice formed over a range of years. The properties of the ice and the recrystallized inclusions within the ice can then be used to reconstruct a climatic record over the age range of the core, normally through isotopic analysis.
Carrying capacity	The carrying capacity of a biological species in an environment is the maximum population size of the species that the environment can sustain indefinitely, given the food, habitat, water and other necessities available in the environment. In population biology, carrying capacity is defined as the environment's maximal load, which is different from the concept of population equilibrium.

For the human population, more complex variables such as sanitation and medical care are sometimes considered as part of the necessary establishment.

Demographic transition	The demographic transition is the transition from high birth and death rates to low birth and death rates as a country develops from a pre-industrial to an industrialized economic system. The theory is based on an interpretation of demographic history developed in 1929 by the American demographer Warren Thompson (1887-1973). Thompson observed changes, or transitions, in birth and death rates in industrialized societies over the previous 200 years.
Isotope	Isotopes are variants of a particular chemical element. While all isotopes of a given element share the same number of protons, each isotope differs from the others in its number of neutrons. The term isotope is formed from the Greek roots isos (?σος 'equal') and topos (τ?πος 'place').
Climate model	Climate models use quantitative methods to simulate the interactions of the atmosphere, oceans, land surface, and ice. They are used for a variety of purposes from study of the dynamics of the climate system to projections of future climate. The most talked-about use of climate models in recent years has been to project temperature changes resulting from increases in atmospheric concentrations of greenhouse gases.
Decomposer	Decomposers (or saprotrophs) are organisms that break down dead or decaying organisms, and in doing so carry out the natural process of decomposition. Like herbivores and predators, decomposers are heterotrophic, meaning that they use organic substrates to get their energy, carbon and nutrients for growth and development. Decomposers use deceased organisms and non-living organic compounds as their food source.
Negative feedback	Negative feedback occurs when the output of a system acts to oppose changes to the input of the system, with the result that the changes are attenuated. If the overall feedback of the system is negative, then the system will tend to be stable.

Overview
In many physical and biological systems, qualitatively different iluences can oppose each other. |

Permafrost	In geology, permafrost is soil at or below the freezing point of water 0 °C (32 °F) for two or more years. Ice is not always present, as may be in the case of nonporous bedrock, but it frequently occurs and it may be in amounts exceeding the potential hydraulic saturation of the ground material. Most permafrost is located in high latitudes (i.e. land close to the North and South poles), but alpine permafrost may exist at high altitudes in much lower latitudes.
Positive feedback	Positive feedback is a process in which the effects of a small disturbance on (a perturbation of) a system include an increase in the magnitude of the perturbation. That is, A produces more of B which in turn produces more of A. In contrast, a system that responds to a perturbation in a way that reduces its effect is said to exhibit negative feedback. These concepts were first recognized as broadly applicable by Norbert Wiener in his 1948 work on cybernetics.
Glacier	A glacier is a large persistent body of ice that forms where the accumulation of snow exceeds its ablation (melting and sublimation) over many years, often centuries. At least 0.1 km² in area and 50 m thick, but often much larger, a glacier slowly deforms and flows due to stresses induced by its weight. Crevasses, seracs, and other distinguishing features of a glacier are due to its flow.
Sea level	Mean sea level is a measure of the average height of the ocean's surface (such as the halfway point between the mean high tide and the mean low tide); used as a standard in reckoning land elevation. MSL also plays an extremely important role in aviation, where standard sea level pressure is used as the measurement datum of altitude at flight levels. Measurement To an operator of a tide gauge, MSL means the 'still water level'--the level of the sea with motions such as wind waves averaged out--averaged over a period of time such that changes in sea level, e.g., due to the tides, also get averaged out.
Heat wave	A heat wave is a prolonged period of excessively hot weather, which may be accompanied by high humidity. There is no universal definition of a heat wave; the term is relative to the usual weather in the area. Temperatures that people from a hotter climate consider normal can be termed a heat wave in a cooler area if they are outside the normal climate pattern for that area.

Chapter 18. Global Change: Climate Alteration and Global Warming

Hurricane	Hurricane! (episode: 1616 (308)) is a Nova episode that aired on November 7, 1989 on PBS. The episode describes the fury of a hurricane and the history of hurricane forecasting. The episode features footage of Hurricane Camille of 1969 and Hurricane Gilbert of 1988 and behind the scenes footage at the National Hurricane Center as forecasters tracked Hurricane Gilbert from its formation to its landfall in northern Mexico. Notable meteorologists, Hugh Willoughby, Bob Sheets (then director of the National Hurricane Center) and Jeff Masters were shown in the episode.
Invasive species	Invasive species, a nomenclature term and categorization phrase used for flora and fauna, and for specific restoration-preservation processes in native habitats, with several definitions. • The first definition, the most used, applies to introduced species (also called 'non-indigenous' or 'non-native') that adversely affect the habitats and bioregions they invade economically, environmentally, and/or ecologically. Such invasive species may be either plants or animals and may drupt by dominating a region, wilderness areas, particular habitats, or wildland-urban interface land from loss of natural controls (such as predators or herbivores).
Coral bleaching	Coral bleaching is the loss of intracellular endosymbionts (Symbiodinium, also known as zooxanthellae) through either expulsion or loss of algal pigmentation. The corals that form the structure of the great reef ecosystems of tropical seas depend upon a symbiotic relationship with unicellular flagellate protozoa that are photosynthetic and live within their tissues. Zooxanthellae give coral its coloration, with the specific color depending on the particular clade.
Fly ash	Fly ash is one of the residues generated in combustion, and comprises the fine particles that rise with the flue gases. Ash which does not rise is termed bottom ash. In an industrial context, fly ash usually refers to ash produced during combustion of coal.
Growing season	In botany, horticulture, and agriculture the growing season is the period of each year when native plants and ornamental plants grow; and when crops can be grown. The growing season is usually determined by climate and elevation, and in horticulture and agriculture the plant-crop selection. Depending on the location, temperature, daylight hours (photoperiod), and rainfall, may all be critical environmental factors.

Gulf Stream	The Gulf Stream, together with its northern extension towards Europe, the North Atlantic Drift, is a powerful, warm, and swift Atlantic ocean current that originates at the tip of Florida, and follows the eastern coastlines of the United States and Newfoundland before crossing the Atlantic Ocean. The process of western intensification causes the Gulf Stream to be a northward accelerating current off the east coast of North America. At about , it splits in two, with the northern stream crossing to northern Europe and the southern stream recirculating off West Africa. The Gulf Stream influences the climate of the east coast of North America from Florida to Newfoundland, and the west coast of Europe. Although there has been recent debate, there is consensus that the climate of Western Europe and Northern Europe is warmer than it would otherwise be due to the North Atlantic drift, one of the branches from the tail of the Gulf Stream.
Thermohaline circulation	The term thermohaline circulation refers to the part of the large-scale ocean circulation that is driven by global density gradients created by surface heat and freshwater fluxes. Wind-driven surface currents (such as the Gulf Stream) head polewards from the equatorial Atlantic Ocean, cooling all the while and eventually sinking at high latitudes (forming North Atlantic Deep Water). This dense water then flows into the ocean basins.
Biodiversity	Biodiversity is the degree of variation of life forms within a given species, ecosystem, biome, or an entire planet. Biodiversity is a measure of the health of ecosystems. Biodiversity is in part a function of climate.
Biodiversity hotspot	A biodiversity hotspot is a biogeographic region with a significant reservoir of biodiversity that is under threat from humans.
	The concept of biodiversity hotspots was originated by Norman Myers in two articles in 'The Environmentalist' (1988 ' 1990), revised after thorough analysis by Myers and others in 'Hotspots: Earth's Biologically Richest and Most Endangered Terrestrial Ecoregions'.
	To qualify as a biodiversity hotspot on Myers 2000 edition of the hotspot-map, a region must meet two strict criteria: it must contain at least 0.5% or 1,500 species of vascular plants as endemics, and it has to have lost at least 70% of its primary vegetation.
Hotspot	The places known as hotspots or hot spots in geology are volcanic regions thought to be fed by underlying mantle that is anomalously hot compared with the mantle elsewhere. They may be on, near to, or far from tectonic plate boundaries. There are two hypotheses to explain them.

Chapter 18. Global Change: Climate Alteration and Global Warming

Scientific method	Scientific method refers to a body of techniques for investigating phenomena, acquiring new knowledge, or correcting and integrating previous knowledge. To be termed scientific, a method of inquiry must be based on gathering empirical and measurable evidence subject to specific principles of reasoning. The Oxford English Dictionary says that scientific method is: 'a method or procedure that has characterized natural science since the 17th century, consisting in systematic observation, measurement, and experiment, and the formulation, testing, and modification of hypotheses.' The chief characteristic which distinguishes a scientific method of inquiry from other methods of acquiring knowledge is that scientists seek to let reality speak for itself, and contradict their theories about it when those theories are incorrect, i. e., falsifiability.
Water pollution	Water pollution is the contamination of water bodies (e.g. lakes, rivers, oceans, aquifers and groundwater). Water pollution occurs when pollutants are discharged directly or indirectly into water bodies without adequate treatment to remove harmful compounds. Water pollution affects plants and organisms living in these bodies of water.
Uncertainty	Uncertainty is a term used in subtly different ways in a number of fields, including physics, philosophy, statistics, economics, finance, insurance, psychology, sociology, engineering, and information science. It applies to predictions of future events, to physical measurements already made, or to the unknown. Concepts

Although the terms are used in various ways among the general public, many specialists in decision theory, statistics and other quantitative fields have defined uncertainty, risk, and their measurement as:

1. Uncertainty: The lack of certainty, A state of having limited knowledge where it is impossible to exactly describe the existing state, a future outcome, or more than one possible outcome.
2. Measurement of Uncertainty: A set of possible states or outcomes where probabilities are assigned to each possible state or outcome - this also includes the application of a probability density function to continuous variables
3. Risk: A state of uncertainty where some possible outcomes have an undesired effect or significant loss.
4. Measurement of Risk: A set of measured uncertainties where some possible outcomes are losses, and the magnitudes of those losses - this also includes loss functions over continuous variables.

Knightian uncertainty.

Kyoto Protocol

The Kyoto Protocol is a protocol to the United Nations Framework Convention on Climate Change (UNFCCC or FCCC), aimed at fighting global warming. The UNFCCC is an international environmental treaty with the goal of achieving the 'stabilisation of greenhouse gas concentrations in the atmosphere at a level that would prevent dangerous anthropogenic interference with the climate system.'

The Protocol was initially adopted on 11 December 1997 in Kyoto, Japan, and entered into force on 16 February 2005. As of September 2011, 191 states have signed and ratified the protocol. The only remaining signatory not to have ratified the protocol is the United States.

Chapter 18. Global Change: Climate Alteration and Global Warming

Carbon sequestration	Carbon sequestration is the capture of carbon dioxide (CO_2) and may refer specifically to: • 'The process of removing carbon from the atmosphere and depositing it in a reservoir.' When carried out deliberately, this may also be referred to as carbon dioxide removal, which is a form of geoengineering. • The process of carbon capture and storage, where carbon dioxide is removed from flue gases, such as on power stations, before being stored in underground reservoirs. • Natural biogeochemical cycling of carbon between the atmosphere and reservoirs, such as by chemical weathering of rocks. Carbon sequestration describes long-term storage of carbon dioxide or other forms of carbon to either mitigate or defer global warming. It has been proposed as a way to slow the atmospheric and marine accumulation of greenhouse gases, which are released by burning fossil fuels. Carbon dioxide is naturally captured from the atmosphere through biological, chemical or physical processes.
Efficiency	Efficiency in general describes the extent to which time or effort is well used for the intended task or purpose. It is often used with the specific purpose of relaying the capability of a specific application of effort to produce a specific outcome effectively with a minimum amount or quantity of waste, expense, or unnecessary effort. 'Efficiency' has widely varying meanings in different disciplines.
Sustainability	Sustainability is the capacity to endure. For humans, sustainability is the long-term maintenance of responsibility, which has environmental, economic, and social dimensions, and encompasses the concept of stewardship, the responsible management of resource use. In ecology, sustainability describes how biological systems remain diverse and productive over time, a necessary precondition for human well-being.

1. The _____, together with its northern extension towards Europe, the North Atlantic Drift, is a powerful, warm, and swift Atlantic ocean current that originates at the tip of Florida, and follows the eastern coastlines of the United States and Newfoundland before crossing the Atlantic Ocean. The process of western intensification causes the _____ to be a northward accelerating current off the east coast of North America. At about , it splits in two, with the northern stream crossing to northern Europe and the southern stream recirculating off West Africa. The _____ influences the climate of the east coast of North America from Florida to Newfoundland, and the west coast of Europe. Although there has been recent debate, there is consensus that the climate of Western Europe and Northern Europe is warmer than it would otherwise be due to the North Atlantic drift, one of the branches from the tail of the _____.

   a. Gulf Stream
   b. Great Stink
   c. Green waste
   d. Hazardous waste

2. An _____ is an ice mass that covers less than 50 000 km² of land area (usually covering a highland area). Masses of ice covering more than 50 000 km² are termed an ice sheet.

   _____s are not constrained by topographical features (i.e., they will lie over the top of mountains) but their dome is usually centred on the highest point of a massif.

   a. Ocean surface topography
   b. Inverted topography
   c. Ice cap
   d. Endangered Species Act

3. An _____ is an animal without a backbone. The group includes 97% of all animal species - all animals except those in the chordate subphylum Vertebrata (fish, amphibians, reptiles, birds, and mammals).

   _____s form a paraphyletic group.

---

a. Invertebrate
b. Ivoechiton
c. Ocellochiton
d. Oldhamia

4. _____ refers to the rising average temperature of Earth's atmosphere and oceans, which began to increase in the late 19th century and is projected to continue rising. Since the early 20th century, Earth's average surface temperature has increased by about 0.8 °C (1.4 °F), with about two thirds of the increase occurring since 1980. Warming of the climate system is unequivocal, and scientists are more than 90% certain that most of it is caused by increasing concentrations of greenhouse gases produced by human activities such as deforestation and the burning of fossil fuels. These findings are recognized by the national science academies of all major industrialized nations.[A]

Climate model projections are summarized in the 2007 Fourth Assessment Report (AR4) by the Intergovernmental Panel on Climate Change (IPCC).

a. Carbon budget
b. Carbon dioxide equivalent
c. Global warming
d. Carbon lock-in

5. _____ is a process in which the effects of a small disturbance on (a perturbation of) a system include an increase in the magnitude of the perturbation. That is, A produces more of B which in turn produces more of A. In contrast, a system that responds to a perturbation in a way that reduces its effect is said to exhibit negative feedback. These concepts were first recognized as broadly applicable by Norbert Wiener in his 1948 work on cybernetics.

a. Principia Cybernetica
b. Positive feedback
c. Ratio Club
d. Robopsychology

---

1. a
2. c
3. a
4. c
5. b

You can take the complete Chapter Practice Test

for Chapter 18. Global Change: Climate Alteration and Global Warming
on all key terms, persons, places, and concepts.

Online 99 Cents

http://www.epub89.16.20190.18.cram101.com/

Use www.Cram101.com for all your study needs

including Cram101's online interactive problem solving labs in chemistry, statistics, mathematics, and more.

	Environmental justice
	Sustainability
	Environmental science
	Kuznets curve
	Ecological economics
	Ecosystem services
	Environmental economics
	Millennium Ecosystem Assessment
	Natural capital
	Montreal Protocol
	Stewardship
	Radiation
	International Monetary Fund
	United Nations Environment Programme
	Silent Spring
	Triple bottom line
	Green belt
	Incineration
	Coral reef

_____ | Methane

_____ | Greenhouse gas

_____ | Endangered Species Act

_____ | Global warming

_____ | Water pollution

_____ | Biodiversity

_____ | Pollution

_____ | Species diversity

_____ | Wastewater

CHAPTER HIGHLIGHTS: KEY TERMS, PEOPLE, PLACES, CONCEPTS
Chapter 19. Working Toward Sustainability: Environmental Economics, Equity, and Policy

287

Environmental justice	Environmental justice is 'the fair treatment and meaningful involvement of all people regardless of race, color, sex, national origin, or income with respect to the development, implementation and enforcement of environmental laws, regulations, and policies.' In the words of Bunyan Bryant, 'Environmental justice is served when people can realize their highest potential.'
	Environmental justice emerged as a concept in the United States in the early 1980s; its proponents generally view the environment as encompassing 'where we live, work, and play' (sometimes 'pray' and 'learn' are also included) and seek to redress inequitable distributions of environmental burdens (pollution, industrial facilities, crime, etc).. Root causes of environmental injustices include 'institutionalized racism; the co-modification of land, water, energy and air; unresponsive, unaccountable government policies and regulation; and lack of resources and power in affected communities.'
	Definition The United States Environmental Protection Agency defines as follows:
	'Environmental Justice is the fair treatment and meaningful involvement of all people regardless of race, color, national origin, or income with respect to the development, implementation, and enforcement of environmental laws, regulations, and policies. EPA has this goal for all communities and persons across this Nation.
Sustainability	Sustainability is the capacity to endure. For humans, sustainability is the long-term maintenance of responsibility, which has environmental, economic, and social dimensions, and encompasses the concept of stewardship, the responsible management of resource use. In ecology, sustainability describes how biological systems remain diverse and productive over time, a necessary precondition for human well-being.
Environmental science	Environmental science is an interdisciplinary academic field that integrat physical and biological scienc, (including but not limited to Ecology, Physics, Chemistry, Biology, Soil Science, Geology, Atmospheric Science and Geography) to the study of the environment, and the solution of environmental problems. Environmental science provid an integrated, quantitative, and interdisciplinary approach to the study of environmental systems.

	Related areas of study include environmental studi and environmental engineering.
Kuznets curve	A Kuznets curve is the graphical representation of Simon Kuznets' hypothesis that as a country develops, there is a natural cycle of economic inequality driven by market forces which at first increases inequality, and then decreases it after a certain average income is attained.
	An example of why this happens is that early in development investment opportunities for those who have money multiply, while wages are held down by an influx of cheap rural labor to the cities. Whereas in mature economies human capital accrual, or an estimate of cost that has been incurred but not yet paid, takes the place of physical capital accrual as the main source of growth, and inequality slows growth by lowering education levels because poor people lack finance for their education in imperfect credit markets.
Ecological economics	Ecological economics is a transdisciplinary field of academic research that aims to address the interdependence and coevolution of human economies and natural ecosystems over time and space. It is distinguished from environmental economics, which is the mainstream economic analysis of the environment, by its treatment of the economy as a subsystem of the ecosystem and its emphasis upon preserving natural capital. One survey of German economists found that ecological and environmental economics are different schools of economic thought, with ecological economists emphasizing 'strong' sustainability and rejecting the proposition that natural capital can be substituted by human-made capital.
Ecosystem services	Humankind benefits from a multitude of resources and processes that are supplied by natural ecosystems. Collectively, these benefits are known as ecosystem services and include products like clean drinking water and processes such as the decomposition of wastes. While scientists and environmentalists have discussed ecosystem services for decades, these services were popularized and their definitions formalized by the United Nations 2004 Millennium Ecosystem Assessment (MA), a four-year study involving more than 1,300 scientists worldwide.
Environmental economics	Environmental economics is a subfield of economics concerned with environmental issues. Quoting from the National Bureau of Economic Research Environmental Economics program:

Environmental economics is distinguished from Ecological economics that emphasizes the economy as a subsystem of the ecosystem with its focus upon preserving natural capital. One survey of German economists found that ecological and environmental economics are different schools of economic thought, with ecological economists emphasizing 'strong' sustainability and rejecting the proposition that natural capital can be substituted by human-made capital.

Millennium Ecosystem Assessment	The Millennium Ecosystem Assessment, released in 2005, is an international synthesis by over 1000 of the world's leading biological scientists that analyses the state of the Earth's ecosystems and provides summaries and guidelines for decision-makers. It concludes that human activity is having a significant and escalating impact on the biodiversity of world ecosystems, reducing both their resilience and biocapacity. The report refers to natural systems as humanity's 'life-support system', providing essential 'ecosystem services'.
Natural capital	Natural capital is the extension of the economic notion of capital (manufactured means of production) to goods and services relating to the natural environment. Natural capital is thus the stock of natural ecosystems that yields a flow of valuable ecosystem goods or services into the future. For example, a stock of trees or fish provides a flow of new trees or fish, a flow which can be indefinitely sustainable.
Montreal Protocol	The Montreal Protocol on Substances That Deplete the Ozone Layer (a protocol to the Vienna Convention for the Protection of the Ozone Layer) is an international treaty designed to protect the ozone layer by phasing out the production of numerous substances believed to be responsible for ozone depletion. The treaty was opened for signature on September 16, 1987, and entered into force on January 1, 1989, followed by a first meeting in Helsinki, May 1989. Since then, it has undergone seven revisions, in 1990 (London), 1991 (Nairobi), 1992 (Copenhagen), 1993 (Bangkok), 1995 (Vienna), 1997 (Montreal), and 1999 (Beijing). It is believed that if the international agreement is adhered to, the ozone layer is expected to recover by 2050. Due to its widespread adoption and implementation it has been hailed as an example of exceptional international co-operation, with Kofi Annan quoted as saying that 'perhaps the single most successful international agreement to date has been the Montreal Protocol'.
Stewardship	Stewardship is an ethic that embodies responsible planning and management of resources. The concept of stewardship has been applied in diverse realms, including with respect to environment, economics, health, property, information, and religion, and is linked to the concept of sustainability. Historically, stewardship was the responsibility given to household servants to bring food and drinks to a castle dining hall.

Radiation	In physics, radiation is a process in which energetic particles or energetic waves travel through a medium or space. Two types of radiation are commonly differentiated in the way they interact with normal chemical matter: ionizing and non-ionizing radiation. The word radiation is often colloquially used in reference to ionizing radiation but the term radiation may correctly also refer to non-ionizing radiation.
International Monetary Fund	The International Monetary Fund is an intergovernmental organization that promotes international economic cooperation, focusing in particular on policies that have an impact on the exchange rate and the balance of payments. The organization's stated objectives are to promote international economic cooperation, international trade, employment, and exchange rate stability, including by making resources available to member countries to meet balance of payments needs. Its headquarters are in Washington, D.C. The International Monetary Fund's relatively high influence in world affairs and development has drawn heavy criticism from some sources.
United Nations Environment Programme	The United Nations Environment Programme coordinates United Nations environmental activities, assisting developing countries in implementing environmentally sound policies and practices. It was founded as a result of the United Nations Conference on the Human Environment in June 1972 and has its headquarters in the Gigiri neighborhood of Nairobi, Kenya. United Nations Environment Programme also has six regional offices and various country offices.
Silent Spring	Silent Spring is a book written by Rachel Carson and published by Houghton Mifflin on 27 September 1962. The book is widely credited with helping launch the environmental movement. The New Yorker started serializing Silent Spring in June 1962, and it was published in book form (with illustrations by Lois and Louis Darling) by Houghton Mifflin later that year. When the book Silent Spring was published, Rachel Carson was already a well-known writer on natural history, but had not previously been a social critic.
Triple bottom line	The triple bottom line captures an expanded spectrum of values and criteria for measuring organizational (and societal) success: economic, ecological, and social. With the ratification of the United Nations and ICLEI standard for urban and community accounting in early 2007, this became the dominant approach to public sector full cost accounting. Similar UN standards apply to natural capital and human capital measurement to assist in measurements required by , e.g. the ecoBudget standard for reporting ecological footprint.

Green belt	A green belt is a policy and land use designation used in land use planning to retain areas of largely undeveloped, wild, or agricultural land surrounding or neighbouring urban areas. Similar concepts are greenways or green wedges which have a linear character and may run through an urban area instead of around it. In essence, a green belt is an invisible line encircling a certain area, preventing development of the area allowing wildlife to return and be established.
Incineration	Incineration is a waste treatment process that involves the combustion of organic substances contained in waste materials. Incineration and other high temperature waste treatment systems are described as 'thermal treatment'. Incineration of waste materials converts the waste into ash, flue gas, and heat.
Coral reef	Coral reefs are underwater structures made from calcium carbonate secreted by corals. Corals are colonies of tiny living animals found in marine waters that contain few nutrients. Most coral reefs are built from stony corals, which in turn consist of polyps that cluster in groups.
Methane	Appendix: extraterrestrial methane Methane has been detected or is believed to exist in several locations of the solar system. In most cases, it is believed to have been created by abiotic processes. Possible exceptions are Mars and Titan.
Greenhouse gas	A greenhouse gas is a gas in an atmosphere that absorbs and emits radiation within the thermal infrared range. This process is the fundamental cause of the greenhouse effect. The primary greenhouse gases in the Earth's atmosphere are water vapor, carbon dioxide, methane, nitrous oxide, and ozone.
Endangered Species Act	The Endangered Species Act of 1973 (Endangered Species Act; 7 U.S.C. § 136, 16 U.S.C. § 1531 et seq). is one of the dozens of United States environmental laws passed in the 1970s. Signed into law by President Richard Nixon on December 28, 1973, it was designed to protect critically imperiled species from extinction as a 'consequence of economic growth and development untempered by adequate concern and conservation.' The Act is administered by two federal agencies, the United States Fish and Wildlife Service (FWS) and the National Oceanic and Atmospheric Administration (NOAA).

Global warming	Global warming refers to the rising average temperature of Earth's atmosphere and oceans, which began to increase in the late 19th century and is projected to continue rising. Since the early 20th century, Earth's average surface temperature has increased by about 0.8 °C (1.4 °F), with about two thirds of the increase occurring since 1980. Warming of the climate system is unequivocal, and scientists are more than 90% certain that most of it is caused by increasing concentrations of greenhouse gases produced by human activities such as deforestation and the burning of fossil fuels. These findings are recognized by the national science academies of all major industrialized nations.[A] Climate model projections are summarized in the 2007 Fourth Assessment Report (AR4) by the Intergovernmental Panel on Climate Change (IPCC).
Water pollution	Water pollution is the contamination of water bodies (e.g. lakes, rivers, oceans, aquifers and groundwater). Water pollution occurs when pollutants are discharged directly or indirectly into water bodies without adequate treatment to remove harmful compounds. Water pollution affects plants and organisms living in these bodies of water.
Biodiversity	Biodiversity is the degree of variation of life forms within a given species, ecosystem, biome, or an entire planet. Biodiversity is a measure of the health of ecosystems. Biodiversity is in part a function of climate.
Pollution	Pollution is the introduction of contaminants into a natural environment that causes instability, disorder, harm or discomfort to the ecosystem i.e. physical systems or living organisms. Pollution can take the form of chemical substances or energy, such as noise, heat or light. Pollutants, the components of pollution, can be either foreign substances/energies or naturally occurring contaminants.
Species diversity	Species diversity is the effective number of different species that are represented in a collection of individuals (a dataset). The effective number of species refers to the number of equally-abundant species needed to obtain the same mean proportional species abundance as that observed in the dataset of interest (where all species may not be equally abundant). Species diversity consists of two components, species richness and species evenness.

| Wastewater | Waste Water is any water that has been adversely affected in quality by anthropogenic influence. It comprises liquid waste discharged by domestic residences, commercial properties, industry, and/or agriculture and can encompass a wide range of potential contaminants and concentrations. In the most common usage, it refers to the municipal wastewater that contains a broad spectrum of contaminants resulting from the mixing of wastewaters from different sources. |

---

1. _____ is the extension of the economic notion of capital (manufactured means of production) to goods and services relating to the natural environment. _____ is thus the stock of natural ecosystems that yields a flow of valuable ecosystem goods or services into the future. For example, a stock of trees or fish provides a flow of new trees or fish, a flow which can be indefinitely sustainable.

   a. Non-Proliferation Trust
   b. Natural capital
   c. Peak coal
   d. Peak oil

2. _____s are underwater structures made from calcium carbonate secreted by corals. Corals are colonies of tiny living animals found in marine waters that contain few nutrients. Most _____s are built from stony corals, which in turn consist of polyps that cluster in groups.

   a. Coral reef
   b. Village Earth
   c. Wood-free paper
   d. Xeriscaping

3. _____ is 'the fair treatment and meaningful involvement of all people regardless of race, color, sex, national origin, or income with respect to the development, implementation and enforcement of environmental laws, regulations, and policies.' In the words of Bunyan Bryant, '_____ is served when people can realize their highest potential.'

   _____ emerged as a concept in the United States in the early 1980s; its proponents generally view the environment as encompassing 'where we live, work, and play' (sometimes 'pray' and 'learn' are also included) and seek to redress inequitable distributions of environmental burdens (pollution, industrial facilities, crime, etc).. Root causes of environmental injustices include 'institutionalized racism; the co-modification of land, water, energy and air; unresponsive, unaccountable government policies and regulation; and lack of resources and power in affected communities.'

   Definition
   The United States Environmental Protection Agency defines as follows:

'_____ is the fair treatment and meaningful involvement of all people regardless of race, color, national origin, or income with respect to the development, implementation, and enforcement of environmental laws, regulations, and policies. EPA has this goal for all communities and persons across this Nation.

    a. Environmental Manager
    b. Environmental justice
    c. Environmental policy
    d. Environmental psychology

4. A _____ is a gas in an atmosphere that absorbs and emits radiation within the thermal infrared range. This process is the fundamental cause of the greenhouse effect. The primary _____es in the Earth's atmosphere are water vapor, carbon dioxide, methane, nitrous oxide, and ozone.

    a. Physical properties of greenhouse gases
    b. Nitrous oxide
    c. Greenhouse gas
    d. R-410A

5. _____ is an interdisciplinary academic field that integrat physical and biological scienc, (including but not limited to Ecology, Physics, Chemistry, Biology, Soil Science, Geology, Atmospheric Science and Geography) to the study of the environment, and the solution of environmental problems. _____ provid an integrated, quantitative, and interdisciplinary approach to the study of environmental systems.

Related areas of study include environmental studi and environmental engineering.

    a. Environmental science
    b. Environmental sociology
    c. Environmental surveying
    d. Environmental terrorism

---

1. b
2. a
3. b
4. c
5. a

You can take the complete Chapter Practice Test

for Chapter 19. Working Toward Sustainability: Environmental Economics, Equity, and Policy
on all key terms, persons, places, and concepts.

Online 99 Cents

http://www.epub89.16.20190.19.cram101.com/

Use www.Cram101.com for all your study needs

including Cram101's online interactive problem solving labs in chemistry, statistics, mathematics, and more.

Other Cram101 e-Books and Tests

Want More?
Cram101.com...

Cram101.com provides the outlines and highlights of your textbooks, just like this e-StudyGuide, but also gives you the **PRACTICE TESTS**, and other exclusive study tools for all of your textbooks.

Learn More. *Just click*
http://www.cram101.com/

CPSIA information can be obtained at www.ICGtesting.com
Printed in the USA
BVOW10s0204280815

415563BV00002B/4/P